Perspectives on the Design and Development of School Mathematics Curricula

Perspectives on the Design and Development of School Mathematics Curricula

edited by

Christian R. Hirsch
Western Michigan University
Kalamazoo, Michigan

NATIONAL COUNCIL OF
TEACHERS OF MATHEMATICS

Library of Congress Cataloging-in-Publication Data

Perspectives on the design and development of school mathematics
curricula / edited by Christian R. Hirsch.
    p. cm.
Includes bibliographical references.
ISBN 978-0-87353-599-1
1. Mathematics—Study and teaching—United States. 2. Curriculum
planning—United States. 3. Education—Standards—United States. I.
Hirsch, Christian R. II. National Council of Teachers of Mathematics.
QA13.P469 2007
510.71—dc22
                        2007012668

This book was prepared with support from the Center for the Study of Mathematics
Curriculum, funded by the National Science Foundation (NSF) under Grant No.
ESI-0333879. Any opinions, findings, and conclusions or recommendations expressed in
this material are those of the authors and do not necessarily reflect the views of the NSF.

The National Council of Teachers of Mathematics is a public voice of mathematics
education, providing vision, leadership, and professional development to support teachers
in ensuring mathematics learning of the highest quality for all students.

Printed in the United States of America

Table of Contents

Introduction

CHRISTIAN R. HIRSCH
Western Michigan University, Kalamazoo, Michigan

Part 1: Design and Development of Grades K–5 *Standards*-Based Curricula

MAX BELL
University of Chicago, Chicago, Illinois

ANDY ISAACS
University of Chicago Center for Elementary Mathematics and Science Education, Chicago, Illinois

SUSAN JO RUSSELL
Education Research Collaborative, TERC, Cambridge, Massachusetts

CATHERINE RANDALL KELSO
University of Illinois at Chicago, Chicago, Illinois

E. PAUL GOLDENBERG
Education Development Center, Newton, Massachusetts

NINA SHTEINGOLD
Education Development Center, Newton, Massachusetts

Part 2: Design and Development of Grades 6–8 *Standards*-Based Curricula

Part 3: Design and Development of Grades 9–12 *Standards*-Based Curricula

Part 4: A Synthesis Perspective

Foreword

This volume presents a historical perspective on what is perhaps a unique effort in curriculum development in this country. The directors or associates for fifteen comprehensive curriculum development projects, fourteen of which were funded by the National Science Foundation, offer perspectives on the design principles that guided their work as well as insights into the challenges they faced and the barriers to their success. This book furnishes useful guidance to future curriculum developers and documents an important historical record of school mathematics.

The National Science Foundation (NSF) has supported two large-scale, high-profile curriculum development efforts in mathematics. The first was embodied in the School Mathematics Study Group (SMSG) and resulted in the "new math" of the 1960s. The second began in the early 1990s and resulted in the materials that are the subject of this book.

This latter effort, like the earlier SMSG effort, was precipitated by a perceived national crisis. For SMSG it was *Sputnik,* and for the more recent effort it was widespread views that schools were not up to the task of educating workers for the demands of changing economic and workplace circumstances. Indeed, beginning in the early to mid 1980s, a series of reports pointed to inadequate preparation in science and mathematics in the schools (e.g., *A Nation at Risk* and *Educating Americans for the 21st Century,* in 1983; *The Second International Mathematics Study* in 1985; among others).

A possible new approach to curriculum, teaching, and assessment was presented in *Curriculum and Evaluation Standards for School Mathematics,* released in 1989 by the National Council of Teachers of Mathematics (NCTM). The *Curriculum and Evaluation Standards* called for more emphasis on problem solving, communication, reasoning, and mathematical connections and provided a blueprint for improving mathematics education in the United States. This was the first attempt to define curriculum goals on a national scale at that level of detail.

A number of factors contributed to NSF's decision to put a special emphasis on, and to commit a substantial amount of money to, the development of mathematics instructional materials that reflected the recommendations of the NCTM *Standards.* Among these factors were the reports on the state of mathematics education, the blueprint for improved learning and teaching provided by the *Standards,* and the realization that teachers could not implement the recommendations of the *Standards* without curriculum models. And a crucial factor in the decision was the desire and need for a vehicle to rebuild the budget for the NSF education directorate after its near shutdown in the early 1980s. The alignment of these factors was highly unusual and may not recur soon.

Three solicitations were issued in rapid succession beginning immediately after the release of the NCTM *Curriculum and Evaluation Standards*. By the end of 1992, thirteen awards had been made, totaling approximately $70 million, to develop mathematics instructional materials that reflected the recommendations of the NCTM *Standards*. Two additional development awards, one for elementary school and one for high school, were made in the early 2000s.

The instructional materials that resulted from these awards are dramatically different from the mathematics textbooks that were in use at the time. The NSF-supported materials are not all alike and span a rather broad spectrum of innovation, compared to traditional texts. But all these new materials contain mathematical content that is different and is presented differently from that in traditional texts, require a dramatically different pedagogical and assessment approach, place a greater cognitive demand on students and teachers, and require different classroom management. These features make the implementation of these materials difficult, and for many school districts this detracted from their appeal.

In view of the barriers to large-scale adoption of materials of this sort, it is unlikely that one would have predicted at the time this effort began that the materials would gain the market penetration they have or that they would have a large impact on teaching and learning. The impact is multifaceted. A large number of studies of student outcomes have been conducted, and these studies almost uniformly point to improved learning and increased interest in mathematics. A remarkable outcome is that at the elementary and middle school levels, the market penetration is quite large, at between 20 percent and 25 percent of the market. Although the market share of the high school materials is not as large, these materials too are used widely. Another interesting development is that many of the more traditional texts produced by commercial publishers are incorporating some of the approaches found in the NSF-supported materials. Perhaps one of the most important outcomes of the publication of the NCTM *Curriculum and Evaluation Standards* and the subsequent NSF support for the development and implementation of instructional materials that reflect the views espoused in that document is that most of the major commercial publishers now strive to include among their offerings at least one "*Standards*-based, NSF-funded" set of materials.

An important factor in this story is that the development work and related efforts were sustained over some fifteen years with considerable financial support, even though the budget for the Instructional Materials Development program began to decline in the mid 1990s. The amount spent on the development of the materials discussed in this volume is approximately $105 million. This includes funding for major revisions of about half of them. NSF also invested approximately $35 million in dissemination and implementation projects in further support for the materials development effort. Other NSF investments contributed to implementation efforts, including the State Systemic Initiative, the Urban & Rural Systemic Initiatives, the Urban Systemic Program, and the Local Systemic Change Program. Substantially more than half of the funding for these systemic projects was devoted to professional development programs for teachers organized around the curriculum materials. However, many of the systemic projects focused on both science and mathematics, which makes it impossible to ascribe a dollar amount to the implementation of the mathematics materials.

I am pleased at this serious attempt to capture the thinking from the leaders of the teams that did the work related to this unusual curriculum development effort. Thanks are due to the authors of the chapters and special thanks and recognition are due to Chris Hirsch for initiating the project and to Barbara Reys and her team at the Show-Me Center for supporting a portion of the work.

John S. (Spud) Bradley
National Science Foundation

Acknowledgments

This book represents a continuing collaboration among teams of mathematics curriculum developers in sharing ideas and discussing issues of common interest that formally began with an NSF-funded Gateways Conference at the University of Montana in 1992. I am grateful to each of the chapter authors for documenting project ideas on the design, development, testing, and implementation of innovative approaches to *Standards*-based mathematics education as well as shaping my own thinking about curriculum over the years.

A special debt of gratitude is owed to the directors of the NSF-funded Dissemination and Implementation Centers—June Mark (K–12 Mathematics Curriculum Center), Barbara Reys (Show-Me Center), Eric Robinson (COMPASS), and Sheila Sconiers (ARC Center)—for their help in organizing and facilitating the Mathematics Curriculum Design and Development presession to the 2005 NSF K–12 Mathematics and Science Curricula and Implementation Centers Conference. That meeting stimulated the creation of the chapters in this volume. Funding from the Show-Me Center for support of the presession is gratefully acknowledged.

The work reported in this book would not have been possible without the sustained major commitment by the National Science Foundation as well as the long-term support and encouragement of the Instructional Materials Development program officers, particularly Spud Bradley.

I would also like to thank Dana Cox, a doctoral fellow in the Center for the Study of Mathematics Curriculum at Western Michigan University, for her thoughtful review of the authors' final manuscripts and helpful editorial comments. Finally, a very special thank-you is extended to Hope Smith, who helped with correspondence with authors, prepared the chapters for press, and provided constant support and encouragement throughout the process.

Christian R. Hirsch
Editor

Introduction

Curriculum Materials Matter

Christian R. Hirsch

Mathematics curriculum—what it should be, how it is organized and sequenced, how it is taught, and what students learn—is the core around which school mathematics programs evolve. It has been shown that mathematics curriculum materials are a strong determinant of what students have the opportunity to learn and what they learn (Begle 1973; Usiskin 1985; Robitaille and Travers 1992; Schmidt, McKnight, and Raizen 1997; Schoenfeld 2002). Their powerful influence is explained by Ball and Cohen (1996, p. 6):

> Unlike frameworks, objectives, assessments, and other mechanisms that seek to guide curriculum, instructional materials are concrete and daily. They are the stuff of lessons and units, of what teachers and students do. . . . Not only are curriculum materials well positioned to influence individual teachers' work but, unlike many other innovations, textbooks are already "scaled up" and part of the routine of schools. They have "reach" in the system.

The last two decades have witnessed significant efforts to improve school mathematics in the United States through the design, development, and testing of innovative comprehensive curriculum materials—materials that are intended to serve as the primary source of instruction in mathematics for at least three full years. The majority of this work was stimulated by the publication of *Curriculum and Evaluation Standards for School Mathematics* (National Council of Teachers of Mathematics [NCTM] 1989) and supported by funding from the National Science Foundation (NSF). These curriculum materials, commonly referred to as "*Standards*-based curricula," offer an approach to mathematics teaching and learning that is qualitatively different from conventional practice in content, priorities, organization, and approaches.

The importance of curriculum materials for *Standards*-oriented classroom practice was underscored in early efforts to document the impact of the NCTM *Standards*. Case studies in the Recognizing and Recording Reform in Mathematics Education Project (Ferrini-Mundy and Schram 1997) conducted prior to the availability of the NSF-funded curricula suggested that pedagogical features of the *Standards* such as an emphasis on cooperative-group learning, writing in the mathematics classroom, or discussion and discourse "were more readily taken up by teachers than some of the mathematics-content features" (Ferrini-Mundy 2000, p. 39). That changed with the emergence of the NSF-funded *Standards*-based curricula.

Goals and Features

Because the NSF-funded curriculum projects used the NCTM *Curriculum and Evaluation Standards* and later *Principles and Standards for School Mathematics* as a basic design framework, the resulting curricula reflect to varying degrees certain common goals and features:

- Updated content to include data analysis, probability, and, in the high school curricula, topics from discrete mathematics
- Focus on "big ideas" across grade levels and multiple representations
- Applications that provide a connection between mathematics and the world in which students live and consider interesting
- Connections among ideas across mathematical strands and grade levels
- Incorporation of technological tools, especially calculators
- Attention to issues of equity and access
- Active engagement of students through investigations of important mathematical ideas and solving more-challenging problems
- Focus on depth over coverage to promote deeper understanding of important mathematical ideas
- Support for teachers to become stimulators and guides of inquiry
- Learning opportunities for teachers through extensive teacher guides and professional development opportunities
- Assessment embedded in the curriculum materials and used to guide instruction

How these goals were addressed and how the features were incorporated in the curriculum materials for the targeted grade levels depended on the orientation and design principles of the individual projects.

Evaluation and research on the effectiveness of many of the *Standards*-based curricula described in this volume are reported in Senk and Thompson (2003). Four of the programs listed in the Table of Contents were unable to contribute to that book. MathScape: Seeing and Thinking Mathematically in the Middle Grades and Mathematics: Modeling Our World did not have sufficient data on student outcomes at that time. Development work on Think Math! and the CME Project was just beginning. On the whole, the reported evaluations were promising. Synthesizing the findings across the elementary, middle, and high school programs, Kilpatrick (2003, p. 472) summarized the results:

> [S]tudents studying from *Standards*-based curricula do as well as students studying from traditional curricula on standardized mathematics tests and other measures of traditional content. They score higher than those who have studied from traditional curricula on tests of newer content and processes highlighted in the *Standards* docu-

ment. These results indicate that *Standards*-based curricula are working in classrooms in ways their designers intended for them to work.

Genesis of This Volume

In an effort to document the design principles and development processes employed by these *Standards*-based curriculum projects and to share the curriculum knowledge generated in the process of creating these new curricula, a special presession to the 2005 NSF K–12 Mathematics and Science Curricula and Implementation Centers Conference was organized. Each curriculum project was invited to prepare a background paper articulating its design principles and development process. Authors were asked to consider the following questions to help guide the development of their papers:

Goals

What were the goals to be achieved in creating the curriculum?

Design Principles

At the program and course levels, what principles were used in making decisions about

- the mathematical content, mathematical processes, and mathematical habits of mind to be included and their priorities;
- how that mathematics would be organized and sequenced;
- the role of technology?

At the unit and lesson levels, what principles were used in making decisions about

- the intended use of the instructional materials;
- the nature and sequencing of instructional tasks;
- support for students' learning with understanding;
- support for teachers' learning;
- the assessment of students' learning?

Development Process

How were the curriculum materials planned, written, tested, and refined? In the process:

- What barriers were encountered to following your design principles and how were they addressed?
- What compromises were required?
- Were changes in plans for pilot and field testing necessary?

Lessons Learned and Implications for the Future

- How, if at all, would you refine your design principles and for what reasons?
- How would you revise the development process?

Draft papers were circulated for review in advance of the presession. The majority of the presession was devoted to interactions among developers about the lessons learned and how those lessons might influence future development efforts. Henry Kepner, a former NSF program officer and president-elect of NCTM, was commissioned to read the papers in advance to offer an outside perspective on the work of the developers. Partly on the basis of these presession discussions, authors subsequently expanded and refined their papers.

This book contains the final edited papers from the Mathematics Curriculum Design and Development presession. The volume is divided into four sections as indicated in the Table of Contents. The first three sections are devoted to the design and development of *Standards*-based curricula for grades K–5, 6–8, and 9–12, respectively. Several of these chapters describe how the curricula have been, or are now being, revised to reflect new school-based knowledge and professional recommendations presented in *Principles and Standards for School Mathematics* (NCTM 2000). The final section is a chapter prepared by John Dossey, who was president of NCTM when work on the *Curriculum and Evaluation Standards for School Mathematics* was initiated. That chapter presents a synthesis of important elements of design and development that seem particularly promising or warrant further study. It also points to challenges and opportunities facing future curriculum development efforts.

Taken together, the chapters in this volume represent an unprecedented effort by the profession and the National Science Foundation to improve school mathematics through the development and implementation of research-based curricula. It is hoped that this book will be useful to faculty and students involved in graduate courses on school mathematics curricula; mathematics coordinators and teachers providing curriculum leadership at the state, district, and building levels; classroom teachers who may be using these curricula or want to understand more about research-based curricula; present and future curriculum developers, including teachers who write instructional materials for their own classes; researchers in mathematics education; and publishers.

REFERENCES

Ball, Deborah Loewenberg, and David K. Cohen. "Reform by the Book: What Is—or Might Be—the Role of Curriculum Materials in Teacher Learning and Instructional Reform?" *Educational Researcher* 25, no. 9 (1996): 6–8, 14.

Begle, E. G. "Some Lessons Learned by SMSG." *Mathematics Teacher* 66 (March 1973): 207–14.

Ferrini-Mundy, Joan. "The Standards Movement in Mathematics Education: Reflections and Hopes." In *Learning Mathematics for a New Century,* 2000 Yearbook of the National Council of Teachers of Mathematics (NCTM), edited by Maurice J. Burke, pp. 37–50. Reston, Va.: NCTM, 2000.

Ferrini-Mundy, Joan, and Thomas Schram, eds. *The Recognizing and Recording Reform in Mathematics Education Project: Insights, Issues, and Implications. Journal for Research in Mathematics Education* Monograph No. 8. Reston, Va.: National Council of Teachers of Mathematics, 1997.

Kilpatrick, Jeremy. "What Works?" In *Standards-Based School Mathematics Curricula: What Are They? What Do Students Learn?,* edited by Sharon L. Senk and Denisse R. Thompson, pp. 471–88. Mahwah, N.J.: Lawrence Erlbaum Associates, 2003.

National Council of Teachers of Mathematics (NCTM). *Curriculum and Evaluation Standards for School Mathematics.* Reston, Va.: NCTM, 1989.

———. *Principles and Standards for School Mathematics.* Reston, Va.: NCTM, 2000.

Robitaille, David F., and Kenneth J. Travers. "International Studies of Achievement in Mathematics." In *Handbook of Research on Mathematics Teaching and Learning,* edited by Douglas A. Grouws, pp. 687–709. Reston, Va.: National Council of Teachers of Mathematics, 1992.

Schmidt, William H., Curtis C. McKnight, and Senta A. Raizen. *A Splintered Vision: An Investigation of U.S. Science and Mathematics Education.* Dordrecht, Netherlands: Kluwer, 1997.

Schoenfeld, Alan H. "Making Mathematics Work for All Children: Issues of Standards, Testing, and Equity." *Educational Researcher* 31, no. 1 (2002): 13–25.

Senk, Sharon L., and Denisse R. Thompson, eds. *Standards-Based School Mathematics Curricula: What Are They? What Do Students Learn?* Mahwah, N.J.: Lawrence Erlbaum Associates, 2003.

Usiskin, Zalman. "We Need Another Revolution in Secondary School Mathematics." In *The Secondary School Mathematics Curriculum,* 1985 Yearbook of the National Council of Teachers of Mathematics (NCTM), edited by Christian R. Hirsch, pp. 1–21. Reston, Va.: NCTM, 1985.

Part 1

Design and Development of Grades K–5 *Standards*-Based Curricula

1

The Case of
Everyday Mathematics

Max Bell
Andy Isaacs

THE roots of the University of Chicago School Mathematics Project's (UCSMP) Everyday Mathematics (EM) curriculum extend back many decades, but for the purposes of this chapter, a good starting point is the early 1970s, when Max Bell began to develop a coherent vision of what school mathematics should be attempting to accomplish. Bell's efforts led to a seminal article, "What Does 'Everyman' Really Need from School Mathematics?" (Bell 1974), variations of which have guided the elementary component of UCSMP from that time forward. Research and development by UCSMP and others throughout the 1970s and early 1980s verified the extent to which the system needed fixing and many insights into how one might proceed.

By the early 1980s, this exploratory work had evolved into a serious commitment to creating what would eventually become EM. At the outset of the curriculum development effort, the authors could see the hazards of becoming lost in the myriad details of any such undertaking. To keep themselves on track, they formulated a set of beliefs and principles for curriculum building (see fig. 1.1) and outlined what the mathematics content of the curriculum should be. A description of the research basis for these formulations can be found in Isaacs, Carroll, and Bell (1998).

Max Bell wrote the first draft of this chapter with contributions from Toni Fleming, Karen Fuson, Kathleen Pitvorec, and Sheila Sconiers; Andy Isaacs wrote the second draft. Research and development before 1983 were supported by the National Institutes of Education, the National Science Foundation, and the Benton Foundation. Support for research and development for *Kindergarten Everyday Mathematics* came mainly from the Amoco Foundation through the University of Chicago School Mathematics Program (UCSMP), which also provided base support for the remainder of the grades K–3 program. Additional funding for development of Everyday Mathematics for first grade was from the GTE Corporation, and funding for the second- and third-grade programs was from Everyday Learning Corporation and anonymous donors. The development of the program for grades 4–6 was made possible by a major grant from the National Science Foundation. The conclusions and opinions are those of the authors and do not necessarily reflect the views of the National Science Foundation or other sources of funding.

Beliefs and Design Principles

- Most children begin school with considerable knowledge and intuition about numbers, measures, and spatial relationships.
- From their own experience, children construct conceptual knowledge, computational strategies, problem-solving methods, and links to the world around them.
- Since children begin school with a great deal of knowledge and intuition on which to build, the curriculum should aim significantly higher than it has traditionally done.
- Teachers, parents, and other adults are a very important part of children's experience.
- Children can work with and enjoy content and ideas not just from arithmetic but also from essentially all of elementary mathematics. Investigations in geometry, data and statistics, and early concepts of algebra should begin in kindergarten and continue with greater sophistication throughout the grades.
- The K–6 curriculum should help children make transitions from intuition and concrete operations to abstractions and symbol-processing skills while at the same time building new intuitions that will mature in the years beyond sixth grade.
- Manipulatives and other representations should be included as important tools.
- Mathematics should be woven into daily classroom routines so that it becomes a way of thinking rather than simply another school subject.
- Reforms often fail because they are not practical in the working lives of teachers. The new curriculum should be manageable and should include suggestions and procedures that make teachers' lives easier, at least in the long run.
- Basic arithmetic skills and quick responses are essential for building number sense, estimation skills, and flexibility in a problem-rich environment. The new curriculum should include practical routines to help build such skills and responses.
- Instruction and curriculum should provide rich contexts and accommodate a variety of learning styles.
- Whole-class discussions, small-group explorations, individual practice, problem-solving activities, and guided instruction are all important in providing a balanced curriculum.
- Assessment should be ongoing and should fit the types of activities in which students are engaged.
- The development of skills and concepts in children works best if spread over relatively long time periods, sometimes over several

(continued)

years, with early introduction, multiple exposures with increasing sophistication, and spaced practice.

- Topics should not be taught in isolation. Concepts and skills should be interwoven over time and in a variety of applications.
- An enriched curriculum for elementary school mathematics should do such things as these:
  - Move from a nearly exclusive emphasis on "naked number" calculations to the development of conceptual understanding and problem-solving skills in arithmetic, data explorations, probability, geometry, algebra, and functions.
  - Link mathematics to everyday situations.
  - Link past experiences to new concepts and provide for ongoing spaced reviews.
  - Make considerable use of partner and small-group activities.
  - Include hands-on activities and explorations throughout the K–6 program.
  - Build "fact power" mainly with concept-oriented activities and games.
  - Encourage the use and sharing of multiple strategies.
  - Provide a wide variety of assessment opportunities.
  - Encourage enhanced home-school partnerships.

Fig. 1.1

Mathematical Content: Strands and Themes

The EM authors decided to organize the mathematics content of the grades K–6 curriculum into strands and themes. The nine strands were Numeration and Order Relations; Operations and Number Systems; Measures; Numbers in Reference Frames; Algorithms and Procedures; Data and Chance; Geometry and Spatial Sense; Patterns, Rules, and Functions; and Algebra and Uses of Variables.

Themes are crosscutting big ideas or "habits of mind" that students need to develop throughout their mathematics learning experiences. The authors decided to develop the following six themes in the curriculum: estimation and number sense, algorithmic and procedural thinking, mental arithmetic skills and reflexes, problem solving and mathematical modeling, multiple representations and methods, and links of mathematics to the everyday world.

The Development Process

The development of EM has depended on research in mathematics education, the authors' experience and professional judgment, and empirical feedback from classroom trials. For the first edition, materials for each grade were drafted, field-tested, and revised before commercial publication. For each subsequent edition, feedback from users of the previous commercial edition guided revisions, and only new material was formally field-tested.

In both the original development and the revision there is a big gap between (*a*) the level of detail provided by research and by professional recommendations such as the 1989 *Curriculum and Evaluation Standards* from the National Council of Teachers of Mathematics (NCTM) and (*b*) the level of detail needed to write materials teachers can use. Both in original development and in revision a great deal of professional judgment in the face of uncertainty is required in order to provide the level of specificity teachers need.

The First Edition

Work on what became *Kindergarten Everyday Mathematics* (KEM) began in 1983 with support from the Amoco Foundation funding of UCSMP. The development process for KEM established a pattern that was later followed in the development of all of grades K–3 EM.

1. The authors drafted an outline of the curriculum for an entire grade and descriptions of all lessons and units.
2. Summer writing groups of authors and teachers converted the outline and descriptions to a complete first-draft teacher's guide and student journals.
3. This draft was tried out in a full year of closely watched field testing.
4. The materials were revised on the basis of feedback from the field test.
5. The revised version was released for publication.

The reception of KEM encouraged the development of *First-Grade Everyday Mathematics* (1EM) a couple of years later (funded by a grant from the GTE Corporation). The success of those two programs, the first publications of the newly formed Everyday Learning Corporation (ELC), encouraged the development of 2EM and 3EM, with minimal funding from UCSMP and ELC, uncompensated authorship, and anonymous donors of funds for field testing and revision. The development effort would have ended then for lack of further funding but for a timely grant from the National Science Foundation (NSF), which funded the development, field testing, and revision of 4EM, 5EM, and 6EM. The development process for grades 4–6 was similar to the grades K–3 process: materials for each grade were drafted, field-tested for a full year, and revised on the basis of feedback from the field test. Commercial publication of 6EM, the last grade, took place in 1996.

Several hundred teachers participated in pilot testing and field testing the first and subsequent editions of EM, and their selfless work and astute comments were crucial in helping make the program usable in the generality of classrooms. Whenever field-test teachers reported that this or that lesson, exercise, or pedagogic device was impractical, it was either revised to their satisfaction or deleted from the program.

The EM development process depended heavily on teachers' feedback and studies of students' achievement. With widespread use of EM, a large number of achievement studies have been conducted on the curriculum. These studies employ a variety of methodologies and fall into several broad categories: those conducted by the UCSMP elementary school component, often as part of field testing; a longitudinal study funded by NSF and conducted by Karen Fuson and a team at Northwestern University; program evaluations carried out by individual school districts; and research carried out by independent scholars at universities and research institutes (cf. Carroll and Isaacs 2003 and UCSMP 2005).

The Second Edition

In the late 1990s, ELC, which had been purchased by the Tribune Company but had remained under its original management, funded a project to produce a second edition of EM. The development of the second edition was influenced by findings from the studies of EM carried out at Northwestern and elsewhere and also by detailed critiques of the materials commissioned especially to provide guidance for the revisions.

The second edition offered an opportunity to correct errors and smooth out inconsistencies in the program and also to make two major improvements: a new component for grades 3–5 and a reorganized lesson format for grades 1–6.

The first edition of 6EM had included a Student Reference Book (SRB), a component that was so popular that SRBs for grades 3, 4, and 5 were included in the second edition. These SRBs contain many worked examples, including illustrations of multiple ways to solve a problem, and clear expositions of much of the mathematics in the program. The SRBs provide a reference so that students can work more independently and parents who are not familiar with the program can help their children. Teachers have also found the SRBs to be a valuable resource for the mathematics in EM.

The other major improvement in the second edition of EM was a redesigned lesson organization. Lessons in the first edition of EM had included so many activities for the teacher to choose from that it wasn't always clear what was central to the lesson and what was peripheral. Teachers reported that lessons in the first edition seemed to "jump around" and that there was too much material and not enough guidance about what was really important.

In the second edition, every regular lesson in grades 1–6 was reorganized to have an introductory segment and three main parts. The introduction, which usually lasts for about ten minutes, includes a short set of mental arithmetic exercises, a review of homework, and a "Math Message," a carefully designed task that introduces the main topic of the lesson. Part 1 of the lesson, which lasts for twenty to thirty minutes, usually develops directly from a class discussion of the Math Message and focuses on a single skill or concept. During Part 1, the teacher is active, explaining concepts, leading class discussions, demonstrating procedures, questioning students, and so on. Part 2 of the lesson includes distributed practice, or practice that is spaced out over time, and is therefore not closely related to Part 1. Part 2, which also lasts for twenty to thirty minutes, does not have to be completed at the same time as Part 1, which offers flexibility for teachers to fit the program into the school day. Some activities in Part 2 may be suitable for homework. Part 3 of the lesson furnishes optional activities designed to reinforce or extend the skills and concepts from Part 1. For example, an activity in Part 3 may present an alternative way of teaching a concept from Part 1. This new organization breaks the lesson into manageable chunks, clarifies which parts of the lesson focus on that day's goals and which parts are long-term distributed practice, and increases flexibility for the teacher to fit the program to his or her students' needs. A summary of the research base for distributed practice can be found in UCSMP (2003).

Other significant changes in the second edition included the reworking or replacement of lessons and activities that were problematic, the rearrangement of certain units to distribute work more evenly throughout the year, and the renovation of certain strands that were problematic. The EM approach to paper-and-pencil computation, for example, was refined in the second edition's revisions.

The treatment of computation in the second edition has, roughly speaking, three stages. During the early phases of learning an operation, students are encouraged to devise their own procedures. Students solve problems involving the operations "from first principles," before they have developed or learned systematic procedures for solving such problems. This helps them understand the operations better, and it also gives them valuable experience solving nonroutine problems.

Later, when students better understand the concept of the operation, several alternative algorithms are examined. Some of these algorithms are methods that children have devised from working with one another; others are introduced by the teacher or the materials. This is what might be called the alternative algorithms stage. Students in the class may be using one or more of several algorithms, some of which they may have worked out on their own, others of which the teacher may have introduced. Students are urged to experiment with several algorithms and to become proficient at using at least one alternative. The first edition stopped here, with the expectation that every child would become proficient at one or more of the several alternatives.

The second edition goes one step further. Each child is expected to master a "focus algorithm" for each operation. The focus algorithms have been carefully chosen to be easy to learn, efficient, and powerful enough for all practical uses. The focus algorithms are broadly similar to the algorithms that have been traditionally taught in the United States but are more conceptually transparent and easier to execute. All students are expected to master the focus algorithms, though they are not required to use them if they have alternatives they prefer. Focus algorithms provide a common ground for further work and offer reliable alternatives for students who have not developed effective procedures of their own.

The Third Edition

In 2004, EM's publisher, Wright Group/McGraw-Hill,[1] funded a project to produce a third edition of EM. As with both previous editions, the third edition

---

1. In 2000, the Tribune Company sold all its educational holdings, including the Everyday Learning Corporation, to McGraw-Hill. For the next several years, EM was published by SRA/McGraw-Hill, but in 2004 a corporate reorganization shifted EM to the Wright Group/McGraw-Hill. Unlike the Tribune Company's approach to EM, which left ELC's original management team largely intact, McGraw-Hill has been much more actively involved with the program.

is based on feedback from users and research. Changes being made for the new edition, which is due for release in early 2007, focus especially on improving how students' learning is assessed, how the program can be adapted for use with a broad range of students, and how technology can be used to best advantage.

Certain of the usual and often productive differences in points of view among publishers, editors, and authors have increased in the third edition of EM. One cause for these increased differences is the success of EM. Since the program has become one of the best-selling elementary school mathematics series, pressures to accommodate market requirements have increased. The EM authors have resisted these pressures when they appear to them to conflict with their principles and beliefs, and this has occasionally led to tension with the publisher, which understandably wants books that are saleable. Another issue that EM's authors and publisher face arises from differences in the pace of development at a university and the speed with which commercial textbook publishers are accustomed to work. In the early years of EM, this was not a problem because EM's publisher, ELC, had been founded by the authors and the university and was willing to work at the authors' pace (within reason). Since ELC was sold first to the Tribune Company and then to McGraw-Hill, the commercial context for the development of EM has become increasingly traditional, which has led to difficulties in scheduling the work. For example, the first edition of EM was developed over a period of approximately ten years, the second edition took about four years, and the third edition has been compressed into about two years. A two-year development schedule is not unusual in the world of commercial textbooks, but it is a challenge for the approach the EM authors favor.

Professional Development

The approach to instruction envisioned by the original EM authors was quite different from that to which many teachers were (and are) accustomed. From the beginning of their work in the mid 1980s, the authors recognized that professional development for the teachers would be essential to the success of the curriculum they envisioned. In order to teach the new curriculum as the authors intended, teachers would need to make a number of significant changes:

- Teachers would need to broaden their view of mathematics. They would need to understand that mathematics is much more than arithmetic, and they would need to become comfortable with activities that link mathematics to its uses.
- Teachers would have to use a rich mathematics vocabulary with their students and help them develop good mathematics communication skills.
- Teachers would have to manage a dramatic increase in the amount of partner and group work.
- Teachers would have to interpret a wide variety of strategies used by their students, recognize correct solutions, teach multiple strategies, and encourage students to share their strategies and solutions.

- Even more than in traditional curricula, teachers would have to make adaptations for a wide range of student understanding at any given time.
- Teachers would have to employ new assessment strategies.

The authors recognized that many teachers would need significant professional development in order to make such changes. Teachers would need an initial day or two of orientation to the curriculum, ongoing support during the first year of use, and long-term, in-depth professional development focusing on deepening their understanding of the mathematics and pedagogy in the program.

The authors' recognition of the importance of professional development to the success of the curriculum had three major consequences. One was that any publisher of the materials would need to provide an unprecedented amount of professional development; the second was that the materials themselves would have to be designed to promote teachers' as well as students' learning; and the third consequence was that a wide variety of professional development options would need to be created to accommodate the widely varying needs of teachers.

The need for a publisher that would offer substantial professional development was one reason the authors, together with the University of Chicago, founded a new company to publish EM. Under the leadership of its founding president, Joanne Schiller, this new company, Everyday Learning Corporation (ELC), organized a broad array of professional development opportunities in the early 1990s, including a nationwide network of teacher-consultants, conferences, videotapes, newsletters, and materials that local leaders could use to conduct mathematics workshops.

ELC's professional development offerings were especially effective at providing an initial orientation to the curriculum. But the authors believed that long-term teacher learning would be required for teachers' teaching to be truly transformed. The professional development materials published by ELC, which were developed by Sheila Sconiers of UCSMP, provided one route for promoting such long-term learning, but workshops based on those materials and conducted by local leaders could reach only a fraction of teachers using the EM materials. To promote in-depth teacher development on a large scale, the authors attempted to design the EM materials themselves to promote teachers' learning. The EM teacher materials address the richer mathematics in the program, ways of linking mathematics to its uses, management, and pedagogy. Teachers who read the teacher materials and are willing to learn along with the children learn mathematics from the materials themselves and also learn to use and trust the reform-oriented pedagogy. Another significant contributor to long-term teacher learning is simply the experience of teaching the program; many teachers have confirmed that they have indeed learned a great deal of mathematics while teaching EM.

Failure to supply adequate long-term and in-depth support to teachers has been a major factor in the failure of attempted reforms of grades K–6 mathematics several times in the past century. We should not let it happen again. It will be challenging to provide in-depth training at the scale and in the variety that teachers need, but it is a challenge we must meet if reforms are to succeed over the long run.

Lessons Learned

In a talk at the 1995 annual meeting of the National Council of Supervisors of Mathematics, UCSMP's director, Zalman Usiskin, described a pattern that many efforts to reform education seem to follow. According to Usiskin (1995), early work by pioneers and their apostles is often followed by acceptance of the reform by a vanguard of teachers and others who are looking for better approaches, which in turn leads to a much broader acceptance of the reform by the educational establishment and, eventually, to the oversimplification and overapplication of the intended reform, with predictably unfortunate results. The EM authors have worked to avoid this pattern and have perhaps succeeded to some extent. Nevertheless, reforming elementary school mathematics instruction is hard, and the authors have learned several lessons over the twenty years or so of EM's existence.

From Existence Proof to Market Leader

EM started as an effort to produce an "existence proof" of what is possible in school mathematics. The goal was to produce materials that *could* be used to improve the mathematics experience of the vast majority of schoolchildren in the United States, but the expectation was not that the materials *would* be so used. Rather, the hope was to produce materials that mainstream textbook publishers would emulate, so that the effects of EM would be primarily indirect. The EM authors hoped that their ideas might be copied and improved on by the large publishers who dominate the school market, as happened in the 1960s with the School Mathematics Study Group (SMSG) secondary school curricula, but never with any reform-oriented grades K–6 curricula. This has happened to some extent, but in the event, EM has turned out to be much more successful in the marketplace than its authors had anticipated.

EM succeeded in the market for several reasons. The skill of EM's publisher in marketing the materials is clearly one important reason for the program's success. The NCTM's *Curriculum and Evaluation Standards for School Mathematics* (1989) was also important because it helped create a demand for materials such as EM. The high quality and evident efficacy of the EM materials have no doubt been crucial to the long-term success of the program: Good marketing and strong demand will not sustain sales of a product that doesn't work. Finally, EM actually appears to have moved the market toward it. What began as a niche product became mainstream by changing what the market wanted.

Articulation

The development of EM took place one grade at a time, with the program for each year building on what was done in previous grades. The authors expected that implementation would be done in the same way, starting children with KEM or perhaps 1EM and moving on one year at a time. But for a variety of reasons many school districts have elected instead to adopt all grade levels simultaneously. When students "drop in" to the program for the first time at a grade beyond kindergarten or first grade, there are articulation and other problems in some or all of the following ways:

- Teachers who are new to EM struggle to manage it with students for whom the approach is much different from that to which they have been accustomed.

- When all grade levels are adopted at the same time, there are progressively larger gaps in students' background knowledge at higher grade levels.

- Students who have been accustomed to dwelling on one skill until it is "mastered" are sometimes frustrated by EM's "spiral" (or rather "helix"). It takes time for students to feel comfortable switching to a new task or topic before they have achieved a sense of "getting" the one at hand and to realize that it is coming up again pretty soon.

- Often students need to be given assistance and extra time to develop skill in working cooperatively.

- Students need much encouragement and practice in order to learn to communicate their strategies and solutions.

- Students may not be used to "thinking" in math class, since in the past they were rewarded mainly for efficiently memorizing and repeating a process.

Many districts that adopt all grade levels at once know the hazards involved, have made extra efforts to make it work, and have been successful as both teachers and students rise to the challenges of a considerably richer curriculum. In such adoptions, it turns out to be essential to give teachers beyond kindergarten or first grade "permission" to cover less than the full EM curriculum—sometimes much less—until they have students with more complete EM backgrounds.

Another articulation issue is making the transition from fifth- or sixth-grade EM to middle school mathematics in sixth or seventh grade. Traditional middle school materials often repeat the same arithmetic-dominated content of the elementary school curriculum. Even UCSMP's own middle school curriculum and the first round of NSF-supported middle school curricula were originally developed on the assumption that entering middle school students would have had only the standard arithmetic curriculum. Curriculum aside, many middle school teachers may not be ready to handle EM students, since many may not have the training, credentials, confidence, or basic knowledge to teach the mathematics for which these children are ready. As it happens, the EM authors believe that at least half of the students completing KEM–6EM can successfully move directly into the first course of secondary school mathematics, especially the reform-oriented secondary school programs. Although there is anecdotal evidence that this is what is happening, this belief has not yet been rigorously tested through systematic research.

Continuing Barriers

EM was the first of the reform curricula to appear for elementary school mathematics and has succeeded to a remarkable extent. At this writing, EM is perhaps the best-selling elementary school mathematics curriculum in the nation, something the authors never imagined could happen when they began. The curriculum is also successful in most adoptions, with many reports of good results with international, national, state, and local tests, and few reports to the contrary. Nevertheless, there remain barriers to successful implementation.

- Implementation is difficult for many teachers. Considerable support for teachers is embedded in the EM materials, and more is provided by UCSMP and Wright Group/McGraw-Hill, but even more is needed.
- Better curriculum materials are essential for improving instruction, but better materials alone do not inevitably lead to better teaching and higher student achievement. Better textbooks need to be supported by policies and systems of assessment that are aligned with reform goals.
- Research into the efficacy of EM is ongoing, but better measures of implementation quality are urgently needed for such research.
- Problems of urban school reform are particularly challenging. EM is currently the official curriculum in several of the largest school systems in the nation, and although results are promising, the hurdles are daunting.

As it became available on a year-by-year basis starting in about 1990, the EM curriculum struck a chord with a relatively adventurous group of teachers who appreciated its enriched content, progressive pedagogy, and varied approaches to student assessment. Even this pioneering group experienced difficulties in moving from familiar to new ground—difficulties that have become more acute as relatively larger numbers of teachers have followed these pioneers in using EM. These difficulties are likely to continue as barriers to the widespread use not only of the UCSMP program but also of all programs that attempt to implement enriched content with relatively unfamiliar methodologies. Success in dealing sensitively with these issues may determine the fate of the reforms. Here are a few of the issues that must be addressed:

- A strong attachment to traditional school mathematics. Transforming the culture of school mathematics is not easy. The algorithms traditionally used in the United States for the basic operations, for example, are as much cultural artifacts as efficient mathematical procedures. Many adults are quite attached to the "basics" they learned when they were in school.
- Old tests for new emphases. Tests that reflect reform aims are urgently needed, but current political pressures are driving the system toward cheaper tests that focus on low-level skills. It is not clear how long reform programs can continue to aim at a range of goals far broader than what is assessed on mandated tests.
- Concept and skill development over time versus immediate mastery. Despite decades of evidence that distributed practice is more effective than massed practice, many teachers and others remain attached to the paradigm of immediate, verifiable, and step-by-step mastery.
- False dichotomies. Mathematics education is plagued by false dichotomies: proficiency at skills versus conceptual understanding, calculators versus paper-and-pencil algorithms, direct teaching versus cooperative learning, "objective" tests versus alternative assessment, and so on. Too

many people believe they must choose just one in each dichotomy; wise teachers and good curricula are more eclectic.

- Bloat. As EM has gone through successive editions, the tendency for the books to grow too long has become increasingly problematic. Many issues must be addressed in any new edition, and the easiest way to do so is often by writing more words. As a result, Samuel Johnson's comment about *Paradise Lost*— "None ever wished it longer than it is"—might well be applied to EM. Trying to keep the books a usable length is challenging.

- Creep. The tendency for a program to lose its distinctive point of view over time, to drift away from the principles that made it work, is possibly an even more insidious issue. The EM authors want a program that is practical for real teachers, but they also want a program that is faithful to their ideals. As the authors try to make the program more usable for teachers, they must be careful not to debase it.

Implications for the Future

Some implications for the future seem clear.

- Much better ways of helping teachers gear up to teach standards-based programs such as EM are needed.

- Computer-based toolkits hold real potential for improving instruction, especially in areas such as geometry, data, and games that build arithmetic reflexes.

- Eventually curricula should be created that have many more implementation options, using subsets of activities from a database of activities from the present print-based, one-size-fits-all EM curriculum and other resources.

- Eventual delivery of reform curricula to users should take place in at least three modes: by print (perhaps customized for printing at the point of use), by mass storage devices (e.g., CD, DVD), and by online services. This will enable many more implementation options and will make revising and updating much easier.

REFERENCES

Bell, Max S. "What Does 'Everyman' Really Need from School Mathematics?" *Mathematics Teacher* 67 (March 1974): 196–202.

Carroll, William M., and Andrew Isaacs. "Achievement of Students Using the University of Chicago School Mathematics Project's *Everyday Mathematics*." In *Standards-Based School Mathematics Curricula: What Are They? What Do Students Learn?* edited by Sharon L. Senk and Denisse R. Thompson, pp. 79–108. Mahwah, N.J.: Lawrence Erlbaum Associates, 2003.

Isaacs, Andrew, William Carroll, and Max Bell. "A Research-Based Curriculum: The Research Foundations of the UCSMP *Everyday Mathematics* Curriculum," 1998. www.everydaymath.uchicago.edu/educators/references.shtml.

National Council of Teachers of Mathematics (NCTM). *Curriculum and Evaluation Standards for School Mathematics.* Reston, Va.: NCTM, 1989.

University of Chicago School Mathematics Project (UCSMP). "Distributed Practice: The Research Base," 2003. www.everydaymath.uchicago.edu/educators/references.shtml.

———. "*Everyday Mathematics* Research Summary," 2005. www.everydaymath. uchicago.edu/educators/references.shtml (May 30, 2005).

Usiskin, Zalman. "The Stages of Change." *NCSM Newsletter* 24, no. 4 (1995): 14–22.

2

The Case of

Investigations in Number, Data, and Space

Susan Jo Russell

THE first edition of the grades K–5 mathematics curriculum, Investigations in Number, Data, and Space, was developed from 1990 to 1998 through a grant from the National Science Foundation to TERC. A second edition, developed with support by the National Science Foundation, TERC, and Scott Foresman, is now available. Although fundamental principles have not changed from one version to the next, the embodiment of the principles has been refined and improved, and the implications of the principles have been revisited, resulting in a deepening of these principles or changes in emphasis in our work. What follows is based on the development of both the first and revised programs.

Goals and Guiding Principles

From the beginning of our work on this curriculum fifteen years ago, we have had major goals focused on both students' and teachers' learning that guide the development and implementation of the curriculum. The curriculum is designed to

- support students to make sense of mathematics and to learn that they can be mathematical thinkers;
- focus on computational fluency with whole numbers as a major goal of the elementary grades;

The work on which this chapter is based was funded in part by the National Science Foundation through Grant No. ESI-0095450 to TERC. Any opinions, findings, conclusions, or recommendations expressed here are those of the author and do not necessarily reflect the views of the National Science Foundation. Parts of this chapter are based on various informational documents developed by the staff of Investigations.

The co–principal investigators of the first edition were Susan Jo Russell and Cornelia Tierney. Karen Economopoulos directed the development of the K–2 components. Collaborators included Douglas Clements (State University of New York at Buffalo) and Michael Battista (then at Kent State University, now at Michigan State University). The development of the second edition was codirected by Susan Jo Russell and Karen Economopoulos.

- provide substantive work in important areas of mathematics—rational numbers, geometry, measurement, data, and early algebra—and connections among them;
- emphasize reasoning about mathematical ideas;
- communicate mathematics content and pedagogy to teachers;
- engage the range of learners in understanding mathematics.

Fundamental to our work are three guiding principles that underlie all our decisions:

1. *Students have mathematical ideas.* Students come to school with ideas about numbers, shapes, measurements, patterns, and data. If given the opportunity to learn in an environment that stresses making sense of mathematics, students build on the ideas they already have and learn about new mathematics they have never encountered. Students learn that they are capable of having mathematical ideas, applying what they know to new situations, and thinking and reasoning about unfamiliar problems.

2. *Teachers are engaged in ongoing learning* about mathematics content and about how students learn mathematics. The curriculum provides material for professional development, to be used by teachers individually or in groups, that supports teachers' continued learning as they use the curriculum over several years. The Investigations curriculum materials are designed as much to be a dialogue with teachers as to be a core of content for students.

3. *Teachers collaborate with the students and curriculum materials* to create the curriculum as enacted in the classroom. The only way for a good curriculum to be used well is for teachers to be active participants in implementing it. Teachers use the curriculum to maintain a clear, focused, and coherent agenda for mathematics teaching. At the same time, they observe and listen carefully to students, try to understand how they are thinking, and make teaching decisions based on these observations.

These goals and principles are the touchstones we go back to again and again. We approach both students and teachers as agents of their own learning.

Design Principles

In this section, seven categories of design principles are described that have emerged through the work on the first and second editions of Investigations in Number, Data, and Space. The first three principles focus on the structure and content of the curriculum materials; the next four focus on how the curriculum supports teachers' learning.

Unit Structure

From the beginning, we structured each grade level in Investigations as a series of curriculum units in order to emphasize depth in mathematical thinking rather than exposure to a series of fragmented topics. In the early years of the curriculum, besides providing coherent units of study for students, this structure also encouraged the use of Investigations curriculum units as replacement units, allowing schools or school systems to experiment with this new approach to the teaching and learning of mathematics in four-to-six-week chunks. Today, the modular structure continues to allow gradual implementation of the curriculum. For example, some schools start with a few units each year at each grade level, then add more each year until they are implementing the complete curriculum.

However, the most important reason for structuring the curriculum in units is to offer several weeks of instruction focused on a few, related mathematical content emphases. This choice is also based on our experience that students in the elementary grades need time to focus on a central idea, or a few related ideas, and to develop and practice these ideas across a variety of activities and contexts over anywhere from ten to twenty-five sequential class sessions. The traditional U.S. curriculum has been characterized as "a mile wide and an inch deep" because it traditionally attempted to "cover" so many different topics, often jumping from topic to topic every few days. In an Investigations curriculum unit, students engage with a set of ideas that build across different activities, contexts, and representations. They spend time solving problems and writing about and discussing their approaches and solutions as they make sense of the mathematical ideas.

Structuring the curriculum as a set of coherent, extended instructional units does not automatically solve the problem of an overly full curriculum, with too many topics and too much content. The pressures of numerous, often conflicting, state standards, and our experience that there is a great deal of engaging mathematics content that *can* be reasonably studied by students of any age, resulted in the inclusion of too many units at some grades, especially grades 3 and 4, in the first edition. This left to school systems and teachers choices about which units to include in their core program. As a result, units that include the most innovative work—the geometry units that involve technology, the units that focus on the mathematics of change—were implemented less often than other units that treat more familiar topics.

In the second edition, this unit structure is more standardized across grades with particular attention to how much mathematics can reasonably be studied in a year. Hard choices have had to be made about what material to include and what to omit. We cannot possibly present all the worthwhile mathematics that might be included at a particular grade level. We serve students better by helping them study *some* mathematics in depth than by trying to cover the entire range of possible mathematical topics. We have learned the importance of including clearer guidelines for teachers about the time to be spent on each session in a unit and on the unit as a whole, combined with information about how and when the ideas in a curriculum unit will be further developed, later in the year or in the next grade, to help teachers understand which aspects of a topic are completed and which will be revisited.

An implementation difficulty implicit in the unit structure is that although this structure supports furnishing the time for students to develop and focus on complex ideas, it requires another type of structure to supply review and practice. For example, fourth graders spend seven weeks on multiplication and division in the fall, then work on other content (addition and subtraction, fractions, geometry) for fifteen weeks before they return to multiplication and division in the spring. In the meantime, they need opportunities to continue working on problems that allow them to think further about multiplication and division and to practice what they have learned. For this reason, the Routines in grades K–2 and the Ten-Minute Math (TMM) activities in grades 3–5 are integral parts of the program. Teachers can tailor the basic Routines or TMM activities to the needs of their students for ongoing development and the practice of ideas. Doing both the work in the units and daily Routines or TMM requires time spent on mathematics of seventy to seventy-five minutes a day—a sixty-minute math session and an additional ten to fifteen minutes for Routines or TMM. Some schools and teachers ignore the Routines and TMM—and then become concerned about the lack of opportunities for ongoing practice. In the second edition, material for the teacher about how to use and modify these activities is more explicit and visible in the text of each session.

In summary, the unit structure of the curriculum offers flexibility in implementation, emphasizes coherence and depth in mathematical study, and is combined with other instructional components that supply review and practice across units.

Choosing Mathematics Content

Developing computational fluency with whole numbers is a linchpin of the elementary school curriculum. This development includes the building blocks of computation: understanding the base-ten number system and its place-value notation; understanding the meaning of the operations and their relationships; knowing the basic addition and multiplication number combinations (the "facts") and their counterparts for subtraction and division; estimating reasonable results; interpreting problems embedded in contexts and applying the operations correctly to these problems; and learning, practicing, and consolidating accurate and efficient strategies for computing. It also includes developing curiosity about numbers and operations, their characteristics and how they work, and learning to articulate, represent, and justify generalizations.

One principle of the Investigations curriculum is that time and focus on the building blocks of computational fluency precedes practice and consolidation. Extended time across several grades is spent on each operation. Consider subtraction. In kindergarten and grade 1, students solve subtraction problems by modeling the action of subtraction. By grade 2, students use the inverse relationship between addition and subtraction to solve problems. During grades 2 and 3, they become fluent with the subtraction "facts," and they model and solve a variety of types of subtraction problems, including comparison and missing-part problems. In grade 3, as students' understanding of the base-ten number system grows, they use their understanding of place value to solve problems with larger numbers and to become more efficient with subtraction algorithms. In grades 3 and 4, they can articulate, represent, and justify important generalizations about subtraction; for example, if you add the same amount to each number in a subtraction expression, the difference does not change, as in the equation $483 - 197 = 486 - 200$. In these grades, as their fluency with subtraction increases, students analyze and compare strategies for solving subtraction problems. Once they have become fluent and flexible using more transparent subtraction methods, they are also in a position to appreciate the shortcut notation of the U.S. "traditional" or "borrowing" algorithm for subtraction, analyze how it works, and compare it to other algorithms.

This account gives only a glimpse of the work that helps students develop an understanding of subtraction. Curriculum design requires thoughtful consideration of the similar complexity of the other arithmetic operations.

During the revision of Investigations, as we strengthened the coherence and rigor of the number and operations strand, we were determined not to sacrifice the time and depth required for careful development of ideas in this strand. Many state and local mathematics frameworks continue to include much more mathematics content than can be reasonably studied in a school year. At one point in our work, an editor with the publisher reported that the publisher was analyzing how our curriculum addressed a total of 1,550 different learning goals compiled from state frameworks for grade 4—more than eight goals for each school day. Maintaining a focus on depth and meaning required us to make difficult decisions in which we navigated the morass of varying state standards while keeping ourselves grounded in the experience of real students in real classrooms. We did not, therefore, attempt to create a comprehensive textbook that meets all states' (or even any one state's) criteria. Rather, we designed a curriculum based on the needs of real students and teachers and on our analysis of core content for the elementary grades. This organization of content results in a more coherent, focused experience for students. Although schools may have to create some of their own additions or modifications in order to comply with all their state standards, we hope that this core of important mathematics will increasingly be the focus of classroom learning time.

In order to give the needed attention to number and operations while holding to our goal of including a manageable amount of material at each grade level, we had to judge critically the amount of time that could be spent on other important mathematical content: geometry, measurement, data, and patterns and functions. As we developed the second edition, we also considered more carefully how work in these other content areas could connect to and support work in number and operations. For example, a greater emphasis on the foundations of algebra across the grades opened up important opportunities to strengthen work with number and operations. By creating a strong, coherent content strand in patterns, functions, and change across grades K–5, we were able to connect the work primary-grade students do with repeating patterns to the later work on functions. The work on functions provided interesting problem contexts in which students' work on ratio and on constant rates of change connects to and supports their work on multiplication. Similarly, geometry and measurement offer contexts in which students revisit multiplication and fractions. Within the number and operations units themselves, the articulation, representation, and justification of general claims about the operations (an aspect of early algebraic thinking) strengthen students' understanding of the operations.

Making choices about content in the Investigations curriculum is based on knowledge from research and practice, including our own extensive field testing.

Our choice, balance, sequence, and pace of content is based on

- the centrality of number and operations for elementary school students;
- the importance of exposing all elementary school students to a range of mathematics content, including work with geometry, measurement, data, patterns, functions, and the foundations of algebra;
- the development of foundational ideas that students need in order to build an understanding of mathematics;
- lessons learned from several years of in-depth work with many diverse students and their teachers.

Context and Representation: Visualizing Mathematics

Using representations and contexts to visualize mathematical relationships is an essential principle of Investigations. Students may first use representations or contexts concretely, drawing or modeling with materials. Later, they incorporate these representations and contexts into mental models that they can call on to visualize the structure of problems and their solutions. Students develop the habit of using representations to think with and to explain their thinking to others.

One of our tasks as curriculum writers is to identify contexts and representations that are useful with a whole class of problems, can be extended to accommodate new kinds of numbers as students expand their understanding of the number system, and do not have characteristics that overwhelm or interfere with the focus on mathematics content. In addition to offering representations and contexts that teachers model and encourage students to use, the Investigations materials also encourage students to develop the habit of creating their own representations. In the first edition, representations and contexts were central. In the second edition, on the basis of feedback from classrooms and current thinking, we have reviewed our use of particular representations and contexts and made careful decisions about which to emphasize, which to eliminate, and which to add.

Basic representations in the Investigations curriculum include connecting cubes, the 100 chart (and its variants, the 300, 1,000, and 10,000 chart), number lines, rectangular arrays, and sets of 2D and 3D shapes. Each representation allows access to certain characteristics, actions, and properties of numbers and operations or of geometric properties and relationships.

Different representations offer different models of the mathematics and access to different mathematical ideas. For example, both place-value models and number lines are useful in students' study of subtraction, but they each allow students to see different aspects of subtraction. A student solving the problem $103 - 37$ might think about subtracting the 37 in parts by visualizing a place-value model of 103 (10 tens and 3 ones), subtracting 3 tens, then 7 ones (which, for ease of subtraction from 103, the student might split into $3 + 4$). Another student might think about creating an easier, equivalent problem: $103 - 37 = 106 - 40$. This student might visualize "sliding" the interval from 37 to 103 along a number line to determine how to change the numbers while preserving the difference between them.

Creating good contexts for mathematical work is extremely important in the elementary grades. Contexts that students can imagine and visualize give access to ways of thinking about the mathematical ideas and relationships they are studying. In order for a context to be useful, it must have two essential characteristics. It must be connected enough to students' experience for them to imagine and represent the actions and relationships. And it must have mathematical power—the context must enable students to work with the mathematical structures and relationships that they are investigating.

A good context can be created from familiar events or from fantasy. What is important is that it allows students access to the mathematics without obscuring the mathematics with too much elaboration. The context sets out a concrete world that students can imaginatively enter to explore mathematical relationships. The particulars of this world are designed to engage students in these relationships rather than in nonmathematical aspects of the context. For example, in one of our patterns and functions units, students work with the context of a Penny Jar that contains some number of pennies (the starting amount), then has a certain number of pennies added to it each day. This is one of the contexts used to engage students in exploring a function—the relationship of the number of days to the total number of pennies—that involves a constant rate of change and a fixed amount. Our experience in using this context is that students' access to directly modeling the situation and their knowledge of similar real-world contexts engages students quickly in the mathematics and helps them visualize the mathematical relationships but is not so elaborate that it obscures or distracts from the mathematics.

Students should use representations judiciously and with purpose. A first grader solving word problems that involve addition and subtraction might model every problem with cubes. Another student in the same class might model one or two problems, then, having visually confirmed the action of the operation, might solve the rest by imagining one quantity and counting on. A third student—or the same student later in the year—might reason about the numbers without using a visible image or model, relying on the mental images he or she has developed.

A principal issue for designing curriculum is the careful choice of representations to be introduced to all students. Different representations offer different models of the mathematics and access to different mathematical ideas. A single representation often does not present enough of the richness of the mathematics, but in order to develop familiarity, fluency, and common language within the classroom community for these representations, we must be strategic about the number of representations offered and what they are. Choosing and developing contexts and representations are crucial aspects of curriculum design, so that teachers

- integrate contexts and representations with mathematical power as common tools to think with in the classroom;
- carefully develop, modify, and elaborate contexts so that they are connected to students' knowledge and experience while maintaining their mathematical characteristics;

- support students as they incorporate these contexts and representations in their own mental repertoire;
- understand how each representation or context can be useful and what mathematical ideas it best illuminates.

Components for Teacher Learning

From the beginning, our intention in developing Investigations has been to create a professional development tool for teachers—a tool that furnishes opportunities for learning about mathematics content, about how students learn, and about effective pedagogy. Our design focuses as much on the teacher as learner as on the student as learner. An effective teacher-curriculum partnership depends on a curriculum that offers ongoing learning opportunities for the teacher (Russell 1997, pp. 248–49):

> We see the best mathematics teaching environment as a partnership between teacher and curriculum. Both teacher and curriculum bring important contributions to this partnership that the other cannot do well. It is not possible for most teachers to write a complete, coherent, mathematically sound curriculum. It is not insulting to teachers as professionals to admit this. Curriculum development, like teaching mathematics, is a job that requires people and resources; it requires a skilled team of mathematics educators spending many thousands of hours writing, thinking, working in classrooms, listening to students and teachers....

> But only the teacher is there in the classroom, observing and trying to understand her students' mathematical thinking. Individual teachers must continually assess and modify their mathematics program for their own classroom. Thus, curriculum is not a recipe.... Rather, it provides both a coherent mathematics program for students . . . and material that supports teachers in making better, more thoughtful, more informed decisions about their students' mathematics learning.

> The link between curriculum and teacher decision-making is a focus on mathematical reasoning. Neither curriculum nor teacher can fully anticipate the complex and idiosyncratic nature of the mathematical thinking that might go on among thirty students in a single classroom during any one mathematics class. However, both teacher and curriculum contribute to a repertoire of knowledge about student thinking that leads to better mathematics teaching and learning.

In each curriculum unit, Teacher Notes focus on important mathematical ideas and how students learn them. Because reasoning, articulating, and justifying ideas is such a central part of the curriculum, Dialogue Boxes provide examples of students' discussion. Each unit also includes a brief essay about the mathematics content, examples of students' work, and analyses of assessments.

On the basis of information from the field over the last ten years on the use of the curriculum and on classroom testing of the second edition, we made more careful choices about the material we include for teachers' learning. Investigations units can be perceived as having too much text—too much to read. However, teachers who continue using the materials over several years find that as they teach a curriculum unit more than once, they gradually read more and more of the material, recognizing its relevance to their teaching, until the mathematics content, sequence, and important aspects of students' thinking become second nature. In the revised design, the actual activity sequence is easier for teachers to follow both as they prepare and as they are teaching. At the same time, we added additional features to support teachers' learning—for example, essays about the opportunities for developing the building blocks of algebra in the number and operations units, Teacher Notes about what proving looks like in the elementary grades, and short essays written by teachers about ways in which they work with the range of learners in their classrooms. The professional development components are graduated in their accessibility for new users of the curriculum. At first, the elements embedded directly into the lessons and those that include students' work and dialogue are most accessible and inviting. Once teachers learn to make use of these elements, they are ready to work with some of the Teacher Notes and sets of essays that bring them more deeply into the mathematics content and into their students' learning.

In supporting the implementation of the revised materials, we intend to think more deeply about how to help school systems take advantage of this graduated accessibility and use the materials effectively for teachers' professional development. In sum, the curriculum is designed as a professional development tool that

- offers support for new and more-experienced users in the areas of mathematics content and students' mathematical thinking;
- enables a partnership between teacher and curriculum in the classroom.

Explicitness and Coherence

As Investigations was implemented in a variety of school systems and as we supported professional development and implementation in a variety of ways, we were often surprised to find that what was clear to us about the mathematical focus of a unit, a session, or an activity and about the connection and sequence of the sessions and activities was not always so clear to teachers. Sometimes this happened because the meanings of words were not shared. For example, a teacher might say, "But there's not enough work on place value," but after a conversation, it might become clear that what this teacher meant by "place value" was "expanded notation," not a deeper understanding of the base-ten system. At other times, a teacher might say, "I didn't understand the point of this activity, so I skipped it," when we considered the activity crucial. Or a discussion in the curriculum, as implemented, might become a listing of students' ideas with little focus or direction. Although "mathematical emphases" were listed at the beginning of every class session in the first edition, the words used in these emphases did not always help teachers understand the session's focus and its place in the sequence of class sessions.

Considering how to make the purpose, flow, and connections in the curriculum more "explicit" was central to the revision process. Design decisions resulting from these discussions include

- more careful choice and wording of the mathematical focus of each Investigation and each session;

- the development of benchmarks for students in each unit;
- a clear statement of the focus for each class discussion;
- a Teacher Note, with examples of students' work, for each assessment, related directly to the benchmarks for the unit, including an analysis of work that meets the benchmark(s), partially meets the benchmark(s), or does not meet the benchmark(s);
- more careful choices about the activities included in each Investigation within a unit.

Explicitness and coherence help teachers to
- understand the mathematical "story line" of each unit;
- make connections between the mathematical focus of the unit and each activity and assessment;
- set mathematical goals for each unit and for the school year.

Clarity and Focus for Classroom Discussion

In a curriculum that stresses students' development, investigation, and articulation of mathematical ideas, learning to communicate about those ideas is essential. Such talk includes both teacher-student and student-student interaction. Through talk as well as through writing, students develop their mathematical reasoning and justify their ideas. Class discussion is one site for this work.

We have become more aware of teachers' need for help in focusing class discussion so that this precious time can be productive and useful for students. There are often several possible avenues to pursue in a class discussion. The authors must make clear the focus of the discussion, clarify the reason for that discussion in the sequence of mathematical development, provide ways to frame the discussion, suggest questions to ask that are generative and provoke students' thinking, suggest ways that students can be prepared to participate in the discussion, define possible outcomes of the discussion, and indicate what ideas are likely *not* to be resolved at the end of the discussion. The more the teacher knows about the mathematics content and what the mathematical learning issues are for students in the class, the better prepared the teacher is to make good decisions about the direction of a discussion. However, many teachers need support in order to learn to facilitate discussions in which the mathematical ideas are pushed forward.

In the development of the second edition, therefore, we worked to make the purpose and course of each discussion clearer. However, in our attempt to do so, we at first overly scripted some of these revised discussions. To some of our experienced field-test teachers, the way we were writing the discussions seemed to take away from them what they had learned to do: listen closely to their students, analyze their ideas, and make decisions about next steps based on their thinking. We took this criticism seriously, and although we knew that these teachers did not necessarily represent the "typical" teacher, we also knew that overly scripting a discussion would not produce the kind of reasoned guidance teachers need in order to learn how to engage their students in mathematical discourse. As a result, we developed a set of guidelines for writing discussions in a way that supplies focus and direction but that is not a step-by-step outline of how a discussion might go. Our intention is to provide a clear statement of the discussion's focus, initial and follow-up questions that can support students' thinking, and a sense of the discussion's direction and outcome, including examples of possible student responses. Our revised Dialogue Boxes also have more commentary to help teachers notice particular aspects of the dialogue.

Classroom discussion is a crucial component of the curriculum and must be supported in the text so that teachers

- focus the discussion on important mathematical ideas;
- ask questions that engage students' thinking;
- know how the discussion connects to the work that precedes and follows it.

Support for the Range of Learners

One of the goals listed at the beginning of this chapter that guides our curriculum-design process is to "engage the range of learners in understanding mathematics." This goal requires us to reflect on whether the curriculum engages all students in significant mathematical work and how we can support teachers in making modifications to meet the needs of all students. In the revision, we designed two new elements that focus on this goal.

First, throughout the units, many of the activities include a section called "Supporting the Range of Learners." These sections offer suggestions gleaned from classroom experience about how to modify or extend activities for the variety of students in the class while maintaining focus on the important mathematical ideas. Second, a set of cases at each grade level, written by teachers and framed with commentary written by staff, provide examples of how teachers think through the issues of diversity in their classrooms. These cases are not meant to be exemplars, nor can they offer ready-made solutions for other classrooms; rather, they show the ways in which teachers consider how the needs and strengths of their students vary, how they pose questions about their own teaching, and how they examine the results of different teaching strategies. The cases are organized into the following categories:

- *Setting Up the Mathematical Community.* Teachers write about how they create a supportive and productive learning environment in their classrooms.
- *Accommodations for Learning.* Teachers focus on specific modifications they make to meet the needs of some of their learners—for example, students who are struggling or students who need more challenge.
- *Language and Representation.* Teachers share how they help students use representations and develop language to investigate and express mathematical ideas.

These supports in the curriculum, based on observations of students in field-test classrooms and on ideas from their teachers, help users of the curriculum to

- facilitate the active participation of all students in mathematics;

- acknowledge and build on the strengths students bring to the classroom from their own communities, cultures, and language;
- respect the intellectual work of all students, including those who are struggling;
- develop modifications when students are having difficulty or need more challenge.

The Development Process

An important principle of development for Investigations in Number, Data, and Space, in both the first and the second editions, is its grounding in direct work with students and teachers. This work includes extensive classroom testing of the curriculum, the documentation of classroom observations, regular interaction of staff and teachers (which is also documented), the collection of students' work, and the synthesis and use of all this information in each stage of development. This aspect of our work is apparent in the extensive, carefully catalogued notebooks for each curriculum unit, which contain field notes from each observed class, notes on discussions with teachers, and a summary of all documentation for that unit, focusing on what was learned (at many levels) and including highlights of classroom dialogue. In addition, extensive files of students' work from participating classrooms are maintained in order to monitor the effectiveness of the materials and to furnish a source of examples of a range of students' work to be included in the unit guides.

In the development of the first edition, we worked primarily with individual teachers scattered in a variety of schools—urban, suburban, and rural. A high level of documentation of classroom work was implemented, but we interacted with the teachers mostly individually. In the development of the second edition, besides individual conversations with teachers from a similar mix of school systems, we met regularly with the teachers as a group—for a week in the summer of 2001, followed by monthly meetings during the three field-test years. During the first of those school years, teachers codeveloped activities with the staff and tried out activities in their classrooms. The second and third school years were full field-test years in which teachers taught the revised curriculum for the entire school year. The three-hour monthly meetings included professional development focused on mathematics content, discussions of classroom issues, and meetings in grade-level groups with staff members. These grade-level meetings were invaluable for our understanding and prioritizing of issues about what was working and what was not. The role of the teachers was central in keeping the staff aware of all the complexities of implementing this curriculum well. In particular, these meetings with teachers, along with the classroom observations, helped staff to make the activities more accessible to more students; find the balance between the focus and coherence of the mathematics agenda for the class on the one hand and listening and responding to students' ideas on the other; and give up what just wasn't working—even when we loved it from the point of view of mathematical content.

Just as the *use* of curriculum is most successful in partnership with a teacher who is engaged in ongoing reflection about students' learning, the curriculum design *process* is also at its best when it relies on a close partnership between teachers and designers. The daily challenge and connection supplied by the students and teachers who worked with us grounded the development process in the real needs of the classroom.

REFERENCE

Russell, Susan Jo. "The Role of Curriculum in Teacher Development." In *Reflecting on Our Work: NSF Teacher Enhancement in K–6 Mathematics,* edited by Susan N. Friel and George W. Bright, pp. 247–54. Lanham, Md.: University Press of America, 1997.

3

The Case of

Math Trailblazers:
A Mathematical Journey
Using Science and Language Arts

Catherine Randall Kelso

To introduce the Math Trailblazers curriculum, the authors sometimes relate a "creation myth." Howard Goldberg, a physics professor and a principal investigator of Math Trailblazers, visited his daughter's elementary school in the mid 1970s. When reviewing the science text, he was dismayed at the number of errors he found. On questioning the teacher, he was appalled to find that she rarely taught science because she had no confidence in her content knowledge. With further interrogation, he was horrified to learn that she was trained as a teacher at his own institution. Professor Goldberg went home that night and began writing a complete science course for preservice elementary school teachers. With the collaboration of the mathematician Philip Wagreich, this course eventually produced the Teaching Integrated Mathematics and Science (TIMS) Laboratory Investigations. The National Science Foundation (NSF) funded successive generations of TIMS labs and professional development projects that served as precursors to Math Trailblazers.

Design Principles

The Development of Curriculum Principles

Early ideas. Although it oversimplifies the process, the "creation myth" explains the influence of science concepts on the design of Math Trailblazers. The original grant proposed three curricular strands: experiments and activities (50 percent of the curriculum), problem solving (25 percent), and computational algorithms (25 percent) (Wagreich and Goldberg 1989). The experiments would come from the preexisting TIMS materials and serve as "a conceptual and organizational framework" for the new curriculum (Isaacs 1994). They would bring "mathematics out of the placid world of the textbook and into the active realm of the physical world" (Wagreich and Goldberg 1989).

The central ideas of the TIMS materials focused on using the method of science to investigate a small, well-chosen set of quantitative variables. Emphasis on five fundamental variables—length, area, volume, mass, and time—provided a spiraling structure for the curriculum. Each variable would be studied at least once in each grade in different contexts and greater complexity over the years. As described by Goldberg and Boulanger (1981), the exploration of these variables introduced the language and measurement of quantitative variables, thus giving learners tools for scientific investigation and communication.

The TIMS Laboratory Method furnished a structure for investigating and understanding the relationships between and among variables. The method has four steps: (1) drawing a picture of the experiment, (2) collecting and organizing data in a table, (3) graphing the data, and (4) analyzing the data through a series of questions. The questions ask students to identify patterns in the data and express those patterns in some mathematical form. Many labs lead to the exploration of functional relationships that allow students to understand and make predictions about the world and even about mathematics itself (Isaacs 1994). The method presented a structure for introducing students to different types of functions, graphs, and notation.

This laboratory approach formed the basis for many pedagogical principles of the curriculum. The method is highly predictable and allows for multiple representations of concepts. Therefore, it offers structure for the development of mathematical vocabulary and classroom management of hands-on activities. Relationships among variables are represented—concretely through the use of objects under investigation as well as in a picture (a drawing of the lab setup with the variables labeled), a table, and a graph. The use of multiple representations enables students of differing ability levels to fully engage in the mathematics, since students can approach the same content in ways they understand and solve problems using their choices of strategies and tools. The real-world contexts incorporated in the labs help students develop number sense as they work with numbers they have generated themselves by counting or measuring—numbers that are thus meaningful to them—numbers that in Hiebert's terms are strongly linked to their referents (Hiebert 1988). As they deal with experimental error, they develop estimation skills and a sense about when numbers are close. As they look for patterns in their tables and graphs, they are making sense of the numbers before them (Isaacs 1994).

The problem-solving strand as described in the original grant proposal included experiments and activities along with pencil-and-paper problems, games, and puzzles. The goal was to "place the children in situations that call upon the techniques and skills that they have learned, but to do so in a totally new context" (Wagreich and Goldberg 1989).

The description of the computation strand in the original proposal was sparse, but it included a list of arithmetic topics to be taught in roughly the same order and grades as in traditional curricula. The major differences between traditional curricula and the proposed curriculum were identified as follows: "(1) Less time will be devoted to these topics; (2) the long division algorithm will not be taught; (3) greater emphasis will be given to dealing with estimation, rounding, significant digits; . . . (4) decimals will be introduced at an earlier stage, since the calculators will often be giving decimal answers" (Wagreich and Goldberg 1989, p. 6).

A major goal of the project was to deepen and broaden the mathematics taught in elementary schools. To meet this goal, the proposal listed twelve content areas to be woven into the strands throughout the curriculum: measurement, graphing, geometry, algebra, computation, ratios and proportions, probability and statistics, patterns and relationships, the use of calculators, the use of computers, coordinates, and percent.

The evolution of Math Trailblazers. After two years of development and testing, both the developers and NSF identified problems with the original conceptualization of the curriculum. Isaacs, Wagreich, and Gartzman, three of the authors, described this process in a journal article in 1997. The problems included having students confront too many new ideas in the course of the experiments. The labs took much longer than expected, and the big ideas were lost as teachers tried to teach each new concept and skill. The mathematics did not always grow in logical ways, and the labs were not always the best contexts for developing the content. "We were opposed to developing lessons for each separate skill that students would need to do an experiment, but it was apparent that skills and concepts could not be learned within the context of laboratory experiments to the extent that we anticipated" (Isaacs, Wagreich, and Gartzman 1997, p. 198).

The response to these issues was to make major changes to the organizational structure and, therefore, to the design of the curriculum. The growth of mathematics concepts and skills—not science—now shaped the framework of the curriculum. In particular, the development of number concepts and operations (including a long-division algorithm) provided the organizing structure. Successive scope-and-sequence and planning documents aligned more closely with the National Council of Teachers of Mathematics (NCTM) *Curriculum and Evaluation Standards for School Mathematics* (NCTM 1989). Laboratory investigations were integrated into the curriculum and adapted to develop specific mathematical ideas. "Thus, science was to fit the mathematics instead of the other way around" (Isaacs, Wagreich, and Gartzman 1997).

A Working Set of Curriculum Principles

To develop the mathematics and pedagogy of the curriculum, the authors relied heavily on the current mathematics education research. (See Isaacs [1994] for an extensive review of much of the literature that influenced the design of the curriculum. The second edition, Math Trailblazers: Teacher Implementation Guide, presents a short summary of the research foundations for teachers and parents.) Over time, authors developed the following working set of principles for the curriculum using the mathematical ideas from the TIMS materials, mathematics education research, results of the field test, and their experiences as classroom teachers and educators.

- *Students should have access to more rigorous mathematics than what is in traditional textbooks. The mathematics described in the NCTM Standards documents is the mathematics that should be taught in schools.* Early in the development process, the authors developed a set of Curriculum Benchmarks, based on the ideas in the NCTM *Standards* that outlined the mathematical content in each strand. Developers reduced the time allotted to traditional arithmetic and allotted more time to measurement, geometry, ratios and proportions, probability and statistics, graphing, algebra, estimation, mental arithmetic, and patterns and relationships.

  In developing the second edition, the authors reviewed the NCTM *Principles and Standards for School Mathematics* (NCTM 2000). To maintain alignment with *Principles and Standards*, the math facts program was reorganized to reflect the new recommendations.

- *Mathematics is a tool for science, other disciplines, and everyday life. Science, other disciplines, and everyday life supply contexts for doing math.* The variables of length, area, volume, mass, and time are studied in eight to ten laboratory investigations in each grade. The complexity of the tasks and investigations increases from year to year and presents a context for the development of mathematical concepts. For example, in first grade, students use plastic links arranged in groups of ten to measure length, providing opportunities for grouping and counting objects by tens. In fourth grade, students use metersticks to measure the distance cars roll down a ramp to the nearest hundredth of a meter. This activity provides a context for using the meterstick as a number line to study decimals.

- *Mathematics is not a set of rules invented by experts to be memorized by everybody else. It is not a static collection of bits and pieces of algorithmic procedures and terminology.* As stated by Romberg and Tufte (1987, p. 71), "The future emphases of instruction must be on the powerful ideas of mathematics, their interrelatedness, and the development of quantitative reasoning." Individual lessons address multiple content standards. For example, an investigation of the relationship between the drop height and the bounce height of a tennis ball requires students to measure the length of the drops and successive bounces, organize the data in a table, graph the data, identify and describe patterns in the data, and use the patterns to make predictions. This single experiment involves measurement, data analysis, number sense, estimation, graphing, multiplication, patterns and functions, and mathematical reasoning. All these topics relate to a single context that unites them, resulting in powerful mathematics.

- *Mathematics is best learned through active involvement in solving real problems.* Problem solving is used as a context for students to learn new concepts and skills. They practice skills as they apply them in diverse and increasingly challenging contexts that can be purely mathematical or set in the real world. These problems encourage the purposeful use of mathematics.

- *Concepts should be presented with different representations.* Children should represent ideas concretely, pictorially, graphically, symbolically, and verbally and be able to move back and forth among the representations. The use of multiple representations provides access to rigorous mathematics for students of varying ability levels. Within a lesson, some students can work directly with manipulatives to solve problems, whereas others can solve the same problems using graphs or symbols.

- *Mathematics instruction should balance conceptual and skill development.* Developing a strong conceptual understanding is emphasized in all content areas. New conceptual understandings are built on existing skills and concepts; these new understandings in turn support the further development of skills and concepts. The promotion of conceptual understanding should be interwoven with the distributed practice of skills and procedures. For example, the treatment of computation proceeds in several stages: developing meaning for the operation, inventing procedures for solving problems, and becoming more efficient at carrying out procedures, so that students understand when to apply an operation and how to use varied computational methods to solve problems, including nonroutine problems. Procedural fluency is based on solid conceptual understandings. Procedures become part of the students' prior knowledge—on which they build more advanced conceptual and procedural understandings.

- *Assessment should reflect the breadth and balance of the curriculum.* Assessment should be integrated into instruction so that assessment tasks are valuable learning experiences that have their own merit. This is in contrast to the "teach and test" cycle often found in more traditional curricula. A balanced assessment program gives each student an opportunity to demonstrate what he or she knows by allowing multiple approaches, covering a wide range of content, allowing access to appropriate tools, and assessing within different contexts. Therefore, assessments include open-response problems that require an explanation of strategies and the justification of solutions, laboratory investigations that necessitate collaboration over an extended period of time, as well as short, independent, paper-and-pencil tasks.

- *The communication of mathematical ideas is a prime goal of the curriculum.* Communication takes place in ordinary language and mathematical symbols. Reading, writing, talking, drawing, and graphing should be built into mathematics class so that students can share strategies, clarify their thinking, justify solutions, and learn from others. Communicating in classrooms requires students to work collaboratively.

- *All students should have the opportunity to learn rigorous mathematics.* Children come to the classroom with varying abilities and diverse backgrounds. Lessons should be designed to include all students. They should see themselves reflected in the illustrations, text, and stories. Rich problem-solving contexts allow students to make connections among real situations, words, pictures, data, graphs, and symbols and to solve problems in ways each child understands.

The developers envisioned students who are independent thinkers, who take responsibility for their own learning, who are confident problem solvers with a variety of strategies, who communicate effectively, and who value working together to solve challenging problems (Isaacs 1994). Many of these habits of mind are made explicit in the published curriculum as teacher and student rubrics that are used to evaluate students' progress on problem-solving tasks.

The Development Process

The Math Trailblazers development process began with the grant submission in 1989 and continues today as we revise professional development modules for the second edition, prepare the third edition for publication, and concurrently conduct research that will shape the fourth edition. The first edition took seven years to complete (copyright 1997). The fourth edition is planned to arrive in classrooms in 2011.

The Development Team

The writing team began with the two principal investigators—one a mathematician, the other a physicist—and a graduate student, who had elementary school teaching experience. It grew to more than twenty members including mathematicians, elementary and secondary school mathematics and science teachers, graduate students, mathematics educators, illustrators, editors, a videographer, a project manager, and others with a combination of these credentials. This diversity of expertise was essential to the development of a curriculum that was mathematically rigorous, pedagogically innovative, practical in the classroom, engaging to children, grammatically correct, within budget, and submitted to the publisher on time. Successful completion of the project also required indefatigable dedication and infinite reserves of energy.

Advisory Board

The creation of an Advisory Board[1] afforded authors opportunities to bounce ideas off a group with an incredible knowledge of mathematics education. Their advice was rarely unanimous, but their (sometimes heated) debates helped the authors understand the range of ideas in the field and offered important feedback from outside the project. Although the committee as a whole met annually, authors consulted several of the members individually to take advantage of their particular expertise.

Writing

Just as the curriculum principles developed over time, so did the writing process. The original plan called for the development of instructional materials for seven grades (K–6) over a five-year period. Materials for each grade would un-

---

1. TIMS Advisory Board members: Tom Post, Mary Lindquist, Naomi Fisher, Eugene Maier, Donald Chambers, Paul Trafton, Tom Berger, Carl Berger, Lourdes Monteagudo, Hugh Burkhardt, Glenda Lappan, and Elizabeth Phillips.

dergo a development phase, pilot, and field test. Relatively quickly, it became clear that the task was too large for the small group of authors and the five-year time frame. As mentioned above, the results of the pilot test of grades 1 and 2 required the project to revise the design of the curriculum. The development process needed to change as well. Several new members joined the staff at that time, and grade-level teams were formed. The time to complete the development process took seven years, two more than in the original timeline.

The expanded development team devised a plan for writing materials for a given grade. This process was used to write the field-test materials and then revise them to produce final edited manuscripts.

The writing process began with a grade-level team member developing a "concept paper" that outlined the content for each unit of the grade. The content was based on previously written materials (pilot or field-test materials, existing planning documents, Curriculum Benchmarks), feedback from the pilot or field test, and a review of appropriate research. The paper was distributed to team members for comments, and an outline of the units for a grade was finalized at a meeting of all team members.

Individual units were assigned, typically to pairs of team members. Often a senior author with a mathematics background was paired with a classroom teacher. The team members again used existing materials, feedback, and research to write a manuscript for both the student and the teacher materials. The entire grade-level team read and commented in writing on the manuscript. Written comments were detailed and included—but were not limited to—correcting typos, revising sentence structure, suggesting changes to improve the classroom management of the lesson, and commenting on the pedagogy or mathematics of the lesson. At a manuscript meeting, the entire team reviewed the comments and resolved any outstanding issues.

On the basis of the written comments and notes from the manuscript meeting, the authors revised the manuscript. The production team then edited the manuscript and prepared it for production after final review by the authors.

Pilot- and Field-Testing

The pilot-test teachers were recruited in the spring of 1991. The goal was to have twelve teachers for each grade in local classrooms. Each teacher received $4,500, and the publisher furnished all the manipulatives. The campus print shop prepared stapled copies of the teacher and student materials that were distributed at monthly feedback meetings. Teachers were to spend one hour a day writing about what happened in their classrooms, marking up student and teacher pages, and submitting samples of students' work. Project staff offered professional development in the summer. They visited classrooms during the school year to assist and to model lessons.

These were ambitious expectations. While trying to pilot-test materials for grades 3 and 4 and revise the materials for grades 1 and 2 at the same time, the writing team ran out of materials for the pilot test. Most teachers supplied some written feedback; others provided none. Most of the comments on the lessons were too general, such as "too hard" or "fine." However, the combination of feedback from the teachers and the classroom observations of the project staff offered enough information for the developers to make significant changes in the next iteration.

The field test included twenty-one schools in eleven districts in five states. The publisher prepared and distributed the spiral-bound written materials and manipulatives. The teachers received one day of professional development at their schools, and the project staff visited each school once to model lessons for teachers and videotape classrooms. Teachers were to spend one hour each week completing feedback forms, marking up pages, and submitting students' work. They offered additional feedback at two meetings in Chicago. It was sometimes difficult to get useful feedback from the teachers, since they had little support and professional development from the project and often did not understand the philosophy.

Despite these problems, information from the tests was invaluable in finding the balance between theory and practice. At a time when the ideas of the NCTM *Curriculum and Evaluation Standards* were relatively new, these teachers gave the project the information that was needed to make the reform ideas work in classrooms. Those who embraced the new methods provided an existence proof that the curriculum could improve mathematics learning. Reluctant teachers underscored the need for well-written, teacher-friendly materials and long-term professional development programs. Samples of students' work from the field test appear in the published materials as exemplars. About 10 percent of the teachers have remained connected to the project in some capacity. Some have become professional developers, and others have allowed us to make videotapes of their classrooms that are now part of professional development modules. Those schools that adopted the curriculum supplied important feedback on the first edition that shaped the changes incorporated in the second edition. Eight of the field-test teachers are participants in the current research study and are furnishing valuable information from schools that have used some version of the curriculum for as many as ten years.

Preliminary evaluation studies conducted during this period provided evidence that the content placement and approaches in the early versions of the materials were effective. (See Burghardt 1994; Perry, Whiteaker, and Waddoups 1996; Whiteaker, Waddoups, and Perry 1994.)

Instructional Materials

"We believe that the standard textbook is the death of mathematics. . . ."

—Wagreich and Goldberg 1989

The intended use of instructional materials changed through the development process, including changes in the second edition. The original proposal called for a student "core" book that would be used as a reference and would be much shorter than a traditional text. Most of the curriculum would be organized around lab and activity sheets that had accompanying teacher pages. In response to feedback from teachers, students, and parents, the authors chose to organize the curriculum using a unit structure with the following characteristics and components:

- Grades 1–5 are organized into sixteen to twenty units. Each unit builds on the content of the previous unit, supplying students with opportunities to develop concepts and skills over time within new and varied contexts.

- The units integrate content from within mathematics and from other disciplines, in particular science and language arts. For example, in second grade, students develop place-value concepts and explore volume as they group and count the number of marshmallows that fill containers of different sizes and shapes (Isaacs and Kelso 1996). In fifth grade, students use data tables, graphs, and symbols to represent ratios and proportions and solve problems related to investigations of speed and density.

- Student materials include a student text (consumable in grades 1–2 and nonconsumable in grades 3–5), an accompanying workbook, and an Adventure Book with seven to ten illustrated stories.

- Teacher materials are organized into individually bound unit guides that give information on teaching the lessons, including problems for review and practice, notes on the mathematics content, and assessments. These Unit Resource Guides are accompanied by a Teacher Implementation Guide that presents scope-and-sequence information along with background information on the mathematics content and research base of the curriculum. The second edition includes a Facts Resource Guide that highlights the math facts activities and assessments embedded in the units and a CD-ROM of selected blackline masters and assessment tools.

On the basis of the project's work on several professional development grants, the authors have produced a series of professional development materials called Teacher Enhancement Resource Modules (TERMs). Using the TERMs, staff developers can put together workshops of varying lengths that address the individual needs of a district or school, including introductions to the curriculum for new users and workshops on content and pedagogy for experienced users. Most include videos of students and teachers in classrooms so that teachers have images of *Standards*-based instruction.

Current and Future Work

The original development process guided the writing and publication of the first and second editions. The project received a five-year award in 2003 from NSF to fund the Math Trailblazers Research and Revision Study. This study is an examination of the curriculum and its impact on teaching and learning. Its purpose is to shape and guide the revision of the curriculum, as well as to contribute to the general knowledge of the impact of comprehensive, *Standards*-based mathematics curricula on elementary schools. It is composed of the following four research studies conducted by members of the development team at the University of Illinois at Chicago and researchers from other universities.

1. The Implementation Study is a systematic investigation of the implementation of the curriculum in grades K–5 classrooms that includes survey

data on lessons and units used (and omitted), interviews with teachers, students' work samples, and classroom observations.

2. The Whole Number Study will analyze data from classroom observations in conjunction with data on students' performance on interview tasks in order to understand the relationship between the fidelity of implementation and students' knowledge.

3. The Video Study will analyze the teaching and learning processes in classrooms through the video of first- and fourth-grade classrooms, interviews with teachers, and the testing of students.

4. The Fraction and Proportionality Study will evaluate students' learning of fractions and proportionality in grades 3–5 using fraction and proportionality assessments.

Preliminary analyses of the data indicate that students in Math Trailblazers classrooms can use multiple strategies and tools to solve problems and successfully communicate their reasoning. Analyses of classroom observations show that in high-fidelity classrooms (those in which instruction aligns closely with the written goals of the lesson), teachers provide opportunities for students' learning that include opportunities to reason about mathematical concepts, describe their reasoning and strategies, interpret other students' reasoning, practice mathematical operations, and explore and select appropriate representations. Students in low-fidelity classrooms are not afforded these opportunities as often. In surveys, teachers reported that lessons were particularly successful more often than they reported that lessons were unsuccessful. At the same time, teachers indicated a need for students to have more experiences with important concepts and skills and more opportunities to develop strategies for using the tools and to make connections among representations.

Authors will use the data from these four studies along with reviews and analyses of the assessment program, the mathematics content, and the readability of the student materials to develop a plan for revising the curriculum as they begin the next iteration of curriculum development. Whereas previous editions were shaped by the existing mathematics education research and results of extensive field testing, future editions will benefit from research that is specific to Math Trailblazers classrooms.

References

Burghardt, Birch. "Results of Summative Evaluation Studies: 1993–1994 Evaluation Report, 1994." Unpublished report to the National Science Foundation.

Goldberg, Howard S., and F. David Boulanger. "Science for Elementary School Teachers: A Quantitative Approach." *American Journal of Physics* 49, no. 2 (1981): 120–24.

Hiebert, James. "A Theory of Developing Competence with Written Mathematical Symbols." *Educational Studies in Mathematics* 19, no. 3 (1988): 333–55.

Isaacs, Andrew. "Whole Number Concepts and Operations in Grades 1 and 2: Curriculum and Rationale." D.A. diss., University of Illinois at Chicago, 1994.

Isaacs, Andrew, and Catherine R. Kelso. "Pictures, Tables, Graphs, and Questions: Statistical Processes." *Teaching Children Mathematics* 2 (February 1996): 340–45.

Isaacs, Andrew C., Philip Wagreich, and Martin Gartzman. "The Quest for Integration: School Mathematics and Science." *American Journal of Education* 106, no. 1 (1997): 179–206.

National Council of Teachers of Mathematics (NCTM). *Curriculum and Evaluation Standards for School Mathematics.* Reston, Va.: NCTM, 1989.

———. *Principles and Standards for School Mathematics.* Reston, Va.: NCTM, 2000.

Perry, Michelle, Mikka Whiteaker, and Greg L. Waddoups. "Students' Participation in a Reform Mathematics Classroom: Learning to Become Mathematicians." In *Effects of Reform Mathematics Curricula on Children's Mathematical Understanding,* edited by Karen Fuson. Symposia conducted at the annual meeting of the American Educational Research Association, New York, 1996.

Romberg, Thomas A., and Frederic W. Tufte. "Mathematics Curriculum Engineering: Some Suggestions from Cognitive Science." In *The Monitoring of School Mathematics: Background Papers,* Vol. 2, edited by Thomas A. Romberg and Deborah M. Stewart, pp. 71–108. Madison, Wis.: Wisconsin Center for Education Research, 1987.

Wagreich, Philip, and Howard Goldberg. "A Modern K–6 Mathematics Curriculum Based on Integrating Mathematics and Science." A proposal to the National Science Foundation, University of Illinois at Chicago, 1989.

Whiteaker, Mikka, Greg L. Waddoups, and Michelle Perry. *Teaching Integrated Mathematics and Science (TIMS) in First- and Fourth-Grade Classrooms: An Initial Evaluation.* Institute for Mathematics and Science Education Research Report Series. Chicago, Ill., 1994.

4

The Case of

Think Math!

E. Paul Goldenberg
Nina Shteingold

THINK Math!, a curriculum supported by the National Science Foundation[1] and Harcourt School Publishers, was born from an idea voiced by Barbara Janson at the fifth of a series of annual meetings held by the NSF for directors of their major curriculum development projects. She posed this challenge: How can the assembled group of educators serve teachers and children in districts that, for whatever reasons, resist the adoption of the reform curricula already developed?

Among the explanations for such resistance is the professional development (PD) required for curricular reform, both in the underlying mathematics and in the pedagogical mind-sets and practices that the new programs presuppose. The PD numbers are staggering. It would be daunting enough to serve just those districts that acknowledge the need for reform but believe themselves unready. Offering special PD settings for the entire population of elementary school teachers—roughly 1.6 million—is unimaginable with current budgets. Our idea was to arrange things so that a large part of teachers' requisite learning could happen *during* the natural course of their daily work, in a "learning by doing" model, while already providing high-quality mathematics to the children. This view accords with our experience and that of others (e.g., throughout, and implicit in, the

The Think Math! curriculum was developed by the Division of Mathematics Learning and Teaching, at Education Development Center, Inc. (EDC), Newton, Massachusetts, and is published by Harcourt School Publishers. Its intellectual roots draw significantly from Math Workshop, which represents the mathematical vision of W. W. Sawyer, the brilliant realization and extension of that vision by Robert Wirtz, and the very fortunate collaboration of those two with Morton Botel and Max Beberman.

1. The writing of this chapter and the work that it describes were funded in part by the National Science Foundation (NSF), grant ESI-0099093 to EDC, with additional support from Harcourt School Publishers. The views expressed here are those of the authors and do not necessarily reflect the views of the NSF.

work of Deborah Schifter—e.g., Schifter et al. 1999, but also cf. Ball and Cohen 1999; Borasi and Fonzi 2002; RAND 2003; Cohen 2004) that PD is most successful when grounded in the classroom and closely tied to the content and approaches they would use with their students.

The dual challenge of resistance to reform (as change requires major professional development) and the impracticality of providing adequate professional development support to so many teachers points to a common need: comprehensive curriculum materials blending initial ease-of-use with a design specifically aimed at helping teachers learn. To reach groups that have been missed, such materials would need to address concerns of districts that worry about traditional mathematical content or that are skeptical (or negative) about certain pedagogical reforms. Such materials must be of immediate value to children, even while their teachers learn more. Providing these materials would also be a *first step* in helping districts take *other* steps toward the improvement of their students' mathematical learning.

Trojan Horse Principle: Our approach was to design a package that, on the surface, was more acceptable in the target classrooms than were other reform materials (at that time), and to use that package as a vehicle for delivering both research-based, standards-based instruction and professional development, a kind of Trojan Horse approach. The ultimate aim was to reach new audiences with high-quality mathematics for children while setting few initial hurdles for teachers, to help students *and* their teachers learn by doing.

To be useful for teachers' development through on-the-job learning in the way that Ma (1999) describes, a new kind of elementary school curriculum is needed—"new texts," as Howe (1999) puts it. It follows that to have a large-scale impact, this new curriculum must have the following characteristics:

- It must be *appealing*. The materials must appeal to a broad range of teachers: whether with a skills-and-facts-bound image of mathematics or with a broader view, whether at ease with new pedagogical practices or not, whether they have developed a "profound understanding of fundamental mathematics" (PUFM) or not.

- It must be *easy to use*. They must be usable by the vast majority of teachers with little up-front PD on special pedagogy, manipulatives, theories, mathematical content, or vocabulary.

- It must *promote PD*. They must help teachers who have not developed PUFM to learn more mathematics as they plan and teach. The materials must make familiar content just different enough to get teachers to take a fresh look at the mathematics they already "know." At the same time, both content and style must feel sensible and down-to-earth enough to be approachable and appealing without prior PD effort.

- It must *teach math*. The materials would have to be demonstrably effective with children.

Influencing new audiences depends not only on the nature of the material as described above but also on the ability of the publisher to reach a large market.

Marketability requires that the product (1) be easily seen to meet the criteria listed above, (2) correlate with state tests, and (3) be purchasable as a complete, comprehensive program.

Generic Principles

We have been logging the history of ideas behind the project—a compilation of reviewers' comments and analyses, e-mails among all members of the project (at EDC and Harcourt), records of phone conversations, FAQs, and so on—and attempting to articulate not just decisions but the principles behind those decisions. A major goal was to understand and bring coherence to our own project, but from the very start there was also the idea that there might be some universals worth articulating—some general principles that could help *anyone* who is designing curriculum.

To give some sense of *how* we collected the observations that we call principles, we'll start with a log written by one of us in early August 2004:

> I've begun to notice places where some of the teacher feedback we get, while important for what we can learn from it, is *not* compatible with what we are doing. There are many ways to teach, even many ways to teach well. It does not follow that *combining* ideas from each method makes anything good at all. Recall Goodrow's (1998) research on *Investigations*: though the pure *Investigations* classroom outperformed the pure traditional classroom on certain measures, they were *both* better than the mixed approach. The teachers who combined the methods were quite reasonable in thinking that they could get the best of both worlds. But they were wrong. The failure was not necessarily that they combined the worst of both. It might just be that each method— "traditional" and "constructivist"—has its own integrity, and destroying that integrity means that what's left doesn't work at all.

Many morals might be derived from this story, including "Don't combine curricula" or "Don't combine philosophies." But these don't stand up well as principles. Goodrow's (1998) result does not necessarily argue for a doctrinaire and dogmatic approach to curriculum writing. We *all* base curricula on ideas that have come before us, and our philosophies are inevitably related to others we've encountered. It is hardly imaginable that we evolve ideas that have no antecedents and quite rare that we use only a single one. If one's philosophy is "Take the best from all the available sources," then such eclecticism does not "combine philosophies" at all, and so the injunction against doing so is no protection against supplementing a basal approach with a curriculum like Investigations or vice versa. Selecting activities and ideas (and whatever else goes into curriculum materials) on the basis of their individual appeal, without being aware of what guides one's selection or what the interaction effects might be, seems thoughtless in the most literal sense of the word. We prefer a different interpretation of Goodrow's research:

> Curriculum needs to be *designed*, not patched together. Whatever principles guide it should guide all parts of it.

This still lets us draw ideas from a variety of sources, but it suggests that one must make such selections very carefully and on the basis of conscious principles, so that one does not destroy coherence or, in fact, get the worst of both worlds.

Principles Principle: A curriculum must *have* principles if it is to be coherent. These principles must be held consciously and clearly articulated.[2]

The Trojan Horse Principle creates pressure to mix very traditional ideas (do what teachers expect) and very revolutionary ideas (so that teachers learn those); meanwhile, we know that simply mixing the two doesn't work. We must somehow create a *design* that accommodates both sensibilities.

Inconsistency Postulate: It is impossible to be completely consistent in the application of a principle.

Systems such as education and law are so complex that situations regularly arise in which the basic principles themselves appear to be in conflict, or one is uncertain which principles apply. Principles also, alas, sometimes butt heads with reality.

One source of inconsistencies in curriculum development arises from the necessity to serve too many masters. For example, we want to have kids figure things out on their own. Yet, for marketing and adoptability, we must put directions on student pages (so that decision-makers understand at a glance what students are asked to do) and "teach problem-solving strategies" explicitly.

Accidental inconsistency also happens entirely within the writing process. Materials are written by more than one writer, and it is almost unimaginable, if the group is large enough, for all writers to share a common understanding, let alone a common *acceptance* of the principles. Moreover, accepting a principle *in principle* is not the same as having internalized it and integrated it into one's belief system. It is well known that teachers' "theories in action" are not always in accord with the beliefs they verbalize. The same applies to writers. Just as teachers' beliefs affect their teaching, writers' beliefs affect their writing. Agreeing to a principle, even believing that one believes it, does not guarantee that it shapes the way one teaches or writes.

Inconsistency Principle: Inconsistencies, like deviations from official school grammar, should not arise through inattention. Being guided closely by principles is important, but there are times when a particular principle *can't* be followed. There are also times when one's best judgment tells one to part from a principle that *could* be followed, just as good writers "break rules" effectively. If one is trying to keep track of one's principles, then it is also important to note that when one principle is *deliberately* abrogated, the choice itself reflects an application of another principle.

Having set out to create a Trojan Horse curriculum—the research-based curriculum for moderately traditional schools—it is no surprise that we get pushed

---

2. Requiring "conscious and clearly articulated" principles avoids tautology. If materials have any discernible character—a consistency of style, nature, or approach that distinguishes them from other materials—it's possible to derive patterns that are, in effect, the underlying design principles whether or not the designers had articulated them or even been consciously aware of them. Our ability to find what "must have been the principles" does not guarantee design coherence.

in both directions (by teachers, research, politics, colleagues, our publisher, even our own staff) to create a mixture between a very traditional curriculum (do what teachers and parents expect) and a very revolutionary one. Compromises—in the sense of regressing to the mean, or being incoherently inconsistent—are harmful and would be hated both by our intended audience and by supporters of reform. We claim that it is possible to find solutions that do *not* compromise either our reform aims or the sensibilities and needs of the classrooms we're trying to serve, solutions that instead *satisfy* both interests. This claim, of course, must be tested. The need to test this claim shapes the research that guides development.

The Compromise Postulate: When conflict arises, compromise (at least in the now common sense of regression toward the mean) is *not* a good solution.

Claim: Principles are not idle philosophy. Principles as seemingly practical as "we will choose to teach traditional content, and teach it well enough to get children safely past the obligatory gauntlet of misguided tests, but we will arrange that content to tell a story about mathematical ways of thinking (a.k.a., habits of mind) instead of arranging it by topics" have an effect on everything, from the design of activities, to crafting the look of a page, to the wording used with teachers.

Goldenberg (1999) gives explicit examples of how such an "airy" principle has effects right down to the craft level. The remainder of this chapter will combine a list of issues in curriculum writing that we assume are general with ones that we realize are specific to *our* curriculum.

Specific Design Principles:
A Theory about How to Achieve Our Goals

These principles—many derived directly or indirectly from Sawyer (2003) or Wirtz et al. (1964)—were the foundation for Think Math! and cannot be assumed to be universal.

Curriculum-as-Teacher-Support Principle: Teachers teach the kids; curriculum materials help the teacher. There are things that one can sensibly do as a teacher in class that one cannot write into a set of curriculum materials. Examples: asking children to explain their thinking (the curriculum doesn't know when they've done any!); being humorous (what can be cute as part of a teacher's personality can be irritating and demeaning in print). The materials must not coerce classroom management (teacher notes may recommend group work, but *student* material cannot).

We must not write for one specific kind of teacher or make assumptions about teachers' teaching and management style. We cannot, for example, imagine ourselves teaching a lesson and then record it. We must not assume any special background in mathematics or educational theory. We must not overprescribe, and yet we also must not leave teachers without support and direction.

The Teacher-Interest/Curiosity/Confidence Principle: Elementary school teachers' gaps in content knowledge are not *their* problem. All kindergarten teachers can count and read and write their numbers—they can be emotionally and intellectually engaged in their children's progress, but there's nothing in the *mathematics* to capture their adult minds. We need to make *math* interesting to them while remaining completely appropriate for the children (and not overwhelming to the teachers). The big goal is to get teachers *curious* about the mathematics and *confident* in *their* ability to see more in, and play with, what they already know. Then (we conjecture) it may be easier for them to become curious about their children's learning as well.

Our only contact with the teacher is through print materials, the curriculum they will use. That medium is not optimal for building curiosity in a person already lacking in confidence. In the right context—a well-run class or workshop, for example—new information can give teachers the kind of added knowledge that ultimately builds confidence. But in our situation, the teachers are teaching, not taking, a class. To serve our goals, the print materials must not treat teachers as mathematically limited. Materials can supply mathematical information for teachers, but they cannot appear to take that role as their purpose.

The Math Principle: Each activity must have some purpose that serves the child's *mathematics*. Mathematical learning comes in stages: first encounters with an idea, then exploration and investigation, then building mastery, then practice, application, and so on.

The Do-Not-Compete Principle (gum-chewing and walking a straight line): In any learning activity, gray cells that *should* be devoted to mathematics should not be competed for by arts-and-crafts or other cognitive, verbal, social, organizational, or motor challenges. Cognitive load should be devoted to the mathematics.

This is a very nonobvious principle and we have *seen* it to be easily exaggerated, but the basic idea is that—especially in the early stages of the learning process, during first encounters and exploration—*no* aspect of the activity should compete with the mathematics. Any other demands (physical skills, cooperation skills, verbal skills, data or time-management skills, and so on) must be low.[3] One consequence is that the need to learn how to communicate mathematical ideas and processes, just like the need to learn how to add fractions, must be pursued only when appropriate—not at all opportunities, but only when its own considerable difficulty isn't compounding the difficulty of learning the mathematical ideas themselves. Similarly, contexts or settings that are themselves novel or complex compete with the mathematics for attention. Thought devoted to understanding the context is taken from understanding the mathematics. Taking this principle seriously constrains the activities we develop at different stages of mathematical learning. Any writing, reading, drawing, coloring, cutting and pasting, or vo-

---

3. Low does not mean nonexistent. Discussion, activities with manipulatives, and so on, *can* help children learn, but one must always ask what additional cognitive load is entailed, how much attention the *activity itself* takes and how much is left for the ideas that the activity is intended to generate, and, more generally, whether *this* activity at *this* time is really doing its job, to help children learn mathematics.

cabulary should be both *essential* to the mathematics and the *easiest route to it.* If cutting and pasting is easier than drawing, that's what should be done. If drawing is easier than cutting and pasting, then *that* is what should be done. The decision must be made empirically.

The Drill (but-Not-Mindless-Drill) Principle: Practice is needed, but *mindless* practice puts minds to sleep, defeating the purpose of practice (and certainly the mindfulness).

An ideal approach to practice is to introduce a new idea in a way that heavily *uses* (and thus provides extensive practice with) the old one. At a late enough stage of learning—for example, practice or review—other nonmathematical skills (verbalization or writing or drawing) can be practiced at the same time.

The Language-Learning Principle: Language is for communication, and children are fine-tuned to learn language by communicating. They are quick at learning both vocabulary and usage through acts of communication, and they are notoriously bad at learning these things through definitions and lessons on the language. All vocabulary that *must* be taught (either because it is educationally important or because tests demand it) will be taught through use in context and in communication.

Teachers, tests, and publishers all focus heavily on vocabulary, especially in geometry. A consequence of the Language-Learning Principle is that when children must learn such terms as *face, edge, vertex, prism,* and so on, they must not simply be talking *about* the terms but *using* the terms as the only clear way to talk about something of interest to them. For example, students build a variety of 3D shapes from nets, shapes that include prisms and pyramids, but also *other* objects that are neither of these. To describe these shapes in order to solve puzzles about them, kids *need* precise language. It turns out that the objects can be classified into three groups by comparing the numbers of vertices and faces. In the course of this investigation, students attend to and describe features of the objects. If students saw less-diverse objects—a cube, a brick shape, and a fairly "vanilla" pyramid—there would be no intrinsic need for the terminology, nothing to do with it but know its meaning, nothing that it helps one *express.*

The Language-Use Principle: Limit the reading and writing required of students to the most essential to avoid presenting an extra burden, beyond the learning of new mathematical ideas, for children whose reading, writing, or English skills are not strong. Supply *teachers* with nonverbal *teaching* strategies that use students' observational, pattern-following skills instead of relying solely on auditory processing. Be thoughtful about when to ask children to explain and when not to. Avoid useless boilerplate text like "Solve these problems" or "Fill in the blanks."

Being able to communicate mathematical ideas is a valuable skill in its own right. Discussing one's ideas with others can help both the speaker and the other discussants sharpen their ideas. Moreover, children's words, along with their actions, help inform the teacher about what is going on in the children's minds, which can help guide pedagogy. But putting ideas into words also presents its own significant challenge for all young children, and it can be a real (and avoidable) barrier for those who have difficulties with English in particular or language in general. Some reading/writing/talking is good. It's easy to have too much! (Issues of language use are further discussed in Goldenberg, Shteingold, and Feurzeig 2003.)

The Computational Fluency Principle: Computational fluency is important, not only for political reasons (parents, tests, etc.) but for children's mathematical learning itself.

Without computational fluency, children spend too much of their effort and attention on calculation to be able to get full benefit from mathematical investigations, problem solving, and application. When the only goal is a numeric answer, a calculator may be an adequate substitute, but when one is searching for a pattern, one must *see* parts of the calculation in one's head before even knowing what precision to ask for on the calculator. By itself, the call for computational fluency does not say anything about the value of traditional American whole-number algorithms: they are neither necessary for computational fluency nor sufficient for computational fluency.

The insufficiency of one algorithm, standard or not, requires the project to get children comfortable with looking at a computation, "sizing it up," and then taking a sensible approach. For problems like $341123541 - 9787955$, many methods might work, but the standard algorithm is an ideal choice and there's no reason to avoid it. For problems like $4000 - 2$ or $6354 - 99$, however, reliance on the standard algorithm does not reflect real computational fluency, not because "kids should know lots of ways" but because fluency suggests efficiency, and the standard algorithm is not *efficient* for such computations.

The Making-Friends-with-Numbers Principle: Beyond being computationally fluent, children need, over time, to become "friendly" with various sets of numbers, so that they recognize them when they appear as parts of a pattern.

There is nothing at all special about the sums 1, $1 + 3$, $1 + 3 + 5$, $1 + 3 + 5 + 7$, and so on, unless the numbers 1, 4, 9, 16, and 25 already have some special meaning. Without that extra knowledge, there is a pattern to the problems, but *no pattern* to the answers. All problems, after all, have answers, and we can construct patterned problems that are of no further interest at all. In order to recognize patterns, children must not only get some experience looking for them but also make friends with some special sets of numbers.

All curricula require familiarity with odds and evens as distinct families of numbers, and most call special attention to the multiples of 5 and 10. A few highlight powers of 10. Students cannot, at sight, recognize *all* members of many other sets—they will all learn the multiples of 3, 4, 7, and so on, but only the first ten or so of these multiples—but by learning to think of each of these as a *special* number family, students get the idea of asking new numbers whether they are members of one or more of the familiar families. Primes form such a set. What sets to include is somewhat arbitrary, but some sets particularly stand out. Square numbers (n^2) are important in recognizing and using patterns in multiplication, and they are a valuable foundation for a large variety of algebraic ideas. Think Math! children begin making friends with these numbers in first grade; they should be "old friends" by fourth grade. Other candidates for important sets are 2^n, n^3, 5^n, 25^n (computations using 25), and triangular numbers. In the lower grades, two important ways to get to know a number are to ask how it can be constructed (as a sum, as a product) and to know what the several numbers are just before multiples of any power of 10.

The Current American Standard Algorithms Principle: For each operation, students will learn an algorithm recognizably close to the current American standard.

We would consider computation to be dysfluent if it relied on *only* one algorithm regardless of the nature of the problem, but we have decided that students should learn *an* algorithm—a sequence of steps that provides the result—that allows the student to solve any arbitrary problem accurately and fluently, and with understanding and confidence. Why must it be "recognizably close" to the current American standard algorithm? Because if we do not teach it in *our* style, most teachers (or parents) will teach it anyway in *their* style.[4] Furthermore, they will worry (and be more likely to undermine the curriculum, even unintentionally) if we do not help them learn the standard algorithms. Moreover, teachers and parents tend not to believe that children "know how to multiply" (or add or subtract) until they have learned the "official way." For example, there were instances where we designed activities to build the conceptual basis for, and structure of, an algorithm. To do our activity, the children had to perform computations that they could do well *without* the formal algorithm, but because *teachers* see the computation as *requiring* the algorithm, they "improve the pedagogy" by teaching the algorithm first! What is remarkable is how many teachers will do this *even if they think the activity we've provided is brilliant and are eager to do it afterward!*

This remained a battle on our project itself! Not only are there pro-algorithm and anti-algorithm sentiments, but it has been hard, at times, for people to handle the subtle distinction between, on the one hand, building concepts so that children can compute—learning the algorithm as a common notation for, and generalization of, what they've already learned—and, on the other hand, building concepts for the sake of teaching the algorithm. We do want the former.

The Don't-Oversimplify Principle: Some attempts to simplify mathematics make it harder.

This is a tricky principle to apply, because what constitutes *over*simplification is not well defined and is, in general, a judgment call based on experience or on a particular way of interpreting observation. But the phenomenon is remarkable— and remarkably common—once one sees it, and it deserves some explanation.

> From reviewer notes: One lesson suggests building models of staircases from 1-inch wooden cubes for a whole-class demonstration. [R1] asks if this makes sense:

---

4. We know of curricula that have not included the algorithms; we even know of schools that have banned the teaching of the algorithms and the use of words like *borrow* and *carry*. We have seen *no* examples where such algorithms and vocabulary have not, as a result, merely gone underground. As Jean Benson, one of the writers on our staff, put it, "Would you want *your* child hearing about it for the first time on the street!"

> R1: It would be easier to picture the stairs concept [with the blocks than, say, with a picture alone], but it may be confusing for them to go from 3D to 2D.

Yes, it *may* be confusing, but there's no evidence for it at all, and it hardly seems likely. Do children confuse pictures of dogs with real dogs? If we avoid everything that "*may* be confusing*," we'll dry out the curriculum completely. In our work, we have taken the following two positions:

- All things are innocent of being confusing until proved guilty (by evidence).
- Things that are *known* to be confusing may or may *not* require changes.

> We must still decide whether the confusion is over some irrelevancy to what is being learned—or a *part* of what is being learned. For example, the switch of attention between the lattice points at the intersections of grid lines and the squares that are formed between the grid lines is extremely difficult for children. Both are used (when we count area, we look at the squares between, and when we measure the perimeter, we look at the grid lines themselves), and it takes time not only to understand the content (e.g., area and perimeter) but also to notice the different place for visual attention. Think Math! varies the arrays that kids see very early on, sometimes showing lines and intersections, sometimes showing just the intersections (dots), and sometimes showing arrays of squares. The first few times kids see arrays of squares, they sometimes count the lines that surround the squares rather than the rows or columns of squares themselves, and vice versa. However, this "confusion" is not best handled by using only one representation.

The effort to assure success by reducing confusion or uncertainty and by avoiding large steps has tempted *many* of us to scaffold, explain, and prepare kids more than is needed. It's easy to fall into that trap because we all know that scaffolding and other such approaches are *completely fine* and *good* devices—at times. But not at *all* times. Knowing when to use them is essential.

> R1: *Before writing* the number sentence, review some of the ideas about number lines such as bigger to the right and smaller to the left. Have them do some basic problems, such as point to the 4 and ask what number is two after, two more than, three before, three less than, et cetera.

Why *before* writing? If kids do get confused, won't it be apparent, in sufficient time, *during?* And if they don't get confused without the introduction, why waste the time and risk demeaning them?

It has long been a tenet of reform that connections, understandings, big ideas, and habits of mind represent mathematics more faithfully. Critics have ridiculed the idea of going for such "higher level" goals when our children can't even add and subtract. But removing those "hard" parts may not be simplifying at all: memorizing a catalog of unrelated facts and procedures is difficult.

Craft Principles

The Trojan Horse Principle and the principles that follow from it have practical consequences. In particular, the ease-of-use and teacher-comfort requirements dictate several of the following:

- Exercise care in asking questions and posing problems.

  a. Avoid questions that are too simple. We all know that questions that are too hard can feel defeating. But so can questions that are too simple. Two mechanisms are at work. First, if the answer seems *too* obvious, people wonder if they are missing something. Second, people feel demeaned by questions that don't assume enough of their knowledge or intellect. "Is this all they think I can do?" It is remarkable how often, in classroom observations, the "I don't get it" response is given because the child is expecting *more* than the question demands. In these situations, "I don't get it" should not be interpreted as "I don't know how to do that" but rather as "I don't know what you're asking me to do because it *can't* be as straightforward as it sounds."

  b. The meaning of the question (the start) should be completely unambiguous and not subject to interpretation (no "open start" questions).

  c. Use "open-middle" questions more than "open-ended" ones. Having a clear goal for the end lets the curriculum guide progress. It also gives students and teacher some guidance about what "being correct" is and lets them set (and achieve) clear learning goals. The open-middle allows for multiple approaches, divergent thinking, and creativity and also helps accommodate and include learners with different learning styles and past performances.

  d. We use Headline Stories as a kind of open-ended situation combining problem posing with problem solving. The Headline Stories present information but no clear question. For example, the mundane word-problem starter "Marie had seven marbles and Tim had three marbles" is typically followed by some question like "How many did they have all together?" By omitting the question, the problem is more like real life. There is a fact to be observed, and many questions one might ask about it. How many more does Marie have than Tim? Could they share those marbles equally? Encouraging children to *find* the mathematical questions to ask—and helping them develop a broad range of such questions—is richer than having them simply interpret and solve one question. It may also help them *learn* to interpret questions when they are asked because they will have made up such questions themselves.

- Give teachers fewer places to look as they plan and teach their lesson. This means limiting the number of resources that the teacher *must* use while providing necessary information up front and furnishing additional background that the teacher *can* find.

- Organize material to enable the teacher to see a (rather) big picture. We were inspired by a preexisting curriculum that did not have chapters/units/modules but organized its content into a yearlong, naturally evolving mathematical story (that is, without discrete changes of topic as might be expected at chapter boundaries). We had to give that idea up and organize the curriculum into chapters, in part just to make a useful table of contents.

- Use mainly one big idea per lesson (no "review this while teaching that," at least explicitly). We did *not* compromise here on the Drill-but-Not-Mindless-Drill Principle, but we gave up on calling this interesting opportunity for teacher PD to teachers' attention. The "objectives" of the lesson do not include all of what is in the lesson, largely so first-time users aren't deciding what's the most important idea but are guided by us. As teachers get to know the curriculum, they can see other layers and teach deeper lessons, but the "first time" teaching will still have a valid focus.

- Rethink some ideas that are appropriate for students but revolutionary for many teachers. For example, some playful logic problems that work beautifully with children—are great fun for them and help develop their verbal abilities (clearly explaining their thinking, proving, and more) as well as their mathematical logic—are just too hard for teachers to manage.

- Include the teacher in *reasons* for pedagogical choices. No "do this in groups" (or individually) without a reason. And no "philosophy." Yet we must balance explaining the reasons for our educational choices with the goal of not overwhelming the teacher, and must balance between general ideas and concrete instructions.

- Help the teacher communicate with parents. This means that we must not use jargon from psychology, pedagogical theory, or even mathematics (except for mathematical terms that the *children* must learn). We should think of the parents as intelligent adults, but ones with *no* background in either mathematics or educational philosophy or psychology, and provide explanations for teachers that are clear enough to use with parents.

- Don't suggest that we know how people will feel. "All you have to do is . . ." or "Then you just . . ." or "You simply add, and . . ." all sound like the task is expected to be easy. What if it doesn't feel easy?

- Avoid questions like "Which way do you find easier?" that are designed to focus attention on the method that we'd *like* students to pick.

- Homework has to be comforting for parents. It is inherently a more "conservative" object than any other part of the program is required to be.

Some Battles We Must Lose

We are required to devote much more time to formal testing than we believe is sound. Measuring progress is good, but it is also good not to badger kids constantly. Without letting state standards hijack the curriculum or override our principles, we are required to let the corresponding tests dictate content. (If we do not, many teachers will choose or be forced to put the curriculum on hold and train students to the test.) We must tolerate a crazy timeline of deliverables (to classrooms and to publishers) that does not permit a spiral development of a curriculum. We must endure the tension between the need to develop top-down and bottom-up. In order to take feedback from piloting into consideration, we cannot have even one year's worth of program predeveloped. We must sensitively adjust as we go along, not being too devoted to any master plan to be able to ditch things, change order, add lessons, change methods (even question principles!) as we observe conditions and situations in the classroom. At the same time, we must have a big-picture view prior to working out details.

The fifty-sovereign-nations approach to education in the United States makes it nearly inevitable that curricula remain a mile wide and consequently an inch deep. We cannot focus on important topics when we are required to touch on every small thing. District benchmark tests, designed to pace their teachers to protect the district from the punitive state tests, sometimes demand that all googolplex of the topics be covered in *each* nine-week cycle. Sadder even than being required to throw in the kitchen sink because some state requires it is the fact that if we determine that kids and their teachers can make *more* progress than the frameworks call for, we are pressed *not* to include the more advanced goals. This way lies madness. Or, at best, mediocrity.

Conclusion

We believe that curriculum construction without conscious attention to principles is risky at best, and have articulated that notion as a set of "generic" principles, ones that don't distinguish among curricula and that are, in some ways, not very edifying except to explain the importance of *having* conscious and carefully thought-out principles. We proposed three other kinds of principles:

1. Goal principles (like, for us, the Trojan Horse Principle): These are highly specific to a project: values, unarguable, personal. Others may have different goals, but science is not the arbiter (except to the extent that it is possible to see if one has *achieved* the goals).

2. Theory principles: Designers draw from the literature, and from their own experience, theories about how to achieve the goals. These are (should be!) empirically testable.

Principles of these two types are tempered by accident, time constraints, and other clashes with reality (*not* principles), but they are also tempered by principles of a third type.

3. Practical, or craft, principles—like adjusting for a kind of teacher, student, marketing or political issue, or testing need that one did not anticipate at the start, but *also* developing editorial ideas (like avoiding questions that are too easy)—that evolve over time as one better understands the realities.

If this *is* a useful taxonomy for principles, then the search for universal principles for curriculum design—a knowledge that might be useful in any curriculum development venture—should begin with the practical craft principles. One would look next at the theory-based principles, which, though matters of science, are still *selected* from among complex and sometimes contradictory theories. Conversely, if one wants to understand what gives a curriculum its own character—what ideas and insights keep it from being generic, or what (beyond add-on features that help publishers compete with one another) distinguishes this curriculum from others—one should start from the other direction, looking first at the goals, then the theory, and largely ignoring craft.

The only comfort in listing the battles we believed we had to lose—except to note that, even there, we were sometimes able to apply our principles to make the best of the reality that faced us—is that these identify places where mathematics education, as a profession, has work to do. Where curriculum design *can't* be based on sound mathematics and the best knowledge we collectively have about students' learning because of political, social, or economic factors, we must change *those* factors.

Finally, curriculum writing is not software engineering. The programming of monstrously complex software can be shared among hundreds of writers by creating a top-down modular design, strict naming conventions, clear programming protocols for data-structure and file sharing, and so on. And it can be largely debugged before use. Not so with curriculum. It, too, can be planned from the top down (though there are disadvantages to doing so), but implementing the plan is less easily controlled. There is no obvious list of *relevant* conventions to control; the stylistic matters one *can* regulate are often about form rather than content and likely to be at odds with good teaching. Each writer leaves a mark on the curriculum. Each revision can alter more than needs to be altered; any change inevitably breaks connections that we don't want broken (or hadn't even noticed), causing more change than intended, and sometimes more than we realize. Each significant revision has an effect, and therefore needs its own research. The result of the final revision often gets inadequate research before publication. In today's climate, schools want evidence that "it works" before trying it, but that evidence can't be generated until schools try it. Curriculum design, like teaching, is as much art as it is theory and craft.

References

Ball, Deborah Loewenberg, and David K. Cohen. "Developing Practice, Developing Practitioners: Toward a Practice-Based Theory of Professional Development." In *Teaching as the Learning Profession: Handbook of Policy and Practice,* edited by Gary Sykes and Linda Darling-Hammond, pp. 3–32. San Francisco: Jossey Bass, 1999.

Borasi, Raffaella, and Judith Fonzi. *Foundations: Professional Development That Supports School Mathematics Reform,* Vol. 3. Arlington, Va.: National Science Foundation, 2002.

Cohen, Sophia. *Teachers' Professional Development and the Elementary Mathematics Classroom: Bringing Understanding to Light.* Mahwah, N.J.: Laurence Erlbaum Associates, 2004.

Goldenberg, E. Paul. "Principles, Art, and Craft in Curriculum Design: The Case of Connected Geometry." *International Journal of Computers for Mathematical Learning* 4 (1999): 191–224.

Goldenberg, E. Paul, Nina Shteingold, and Nannette Feurzeig. "Mathematical Habits of Mind for Young Children." In *Teaching Mathematics through Problem Solving: Pre-kindergarten–Grade 6,* edited by Frank K. Lester, Jr., pp. 15–29. Reston, Va.: National Council of Teachers of Mathematics, 2003.

Goodrow, Anne M. "Children's Construction of Number Sense in Traditional, Constructivist, and Mixed Classrooms." Ph.D. diss., Tufts University, 1998.

Howe, Roger. "Book Review: *Knowing and Teaching Elementary Mathematics.*" *Notices of the AMS* 46 (September 1999): 881–87.

Ma, Liping. *Knowing and Teaching Elementary Mathematics: Teachers' Understanding of Fundamental Mathematics in China and the United States.* Mahwah, N.J.: Lawrence Erlbaum Associates, 1999.

RAND Mathematics Study Panel. "Mathematical Proficiency for All Students: Toward a Strategic Research and Development Program in Mathematics Education." RAND Mathematics Study Panel, 2003.

Sawyer, Walter Warwick. *Vision in Elementary Mathematics.* 1964. Reprint. Mineola, N.Y.: Dover Publications, 2003.

Schifter, Deborah, Virginia Bastable, and Susan Jo Russell. *Building a System of Tens, Casebook.* Parsippany, N.J.: Dale Seymour, 1999.

Wirtz, Robert, Morton Botel, Max Beberman, and Walter W. Sawyer. *Math Workshop.* Chicago: Encyclopædia Britannica Press, 1964.

Part 2

Design and Development of Grades 6–8 *Standards*-Based Curricula

5

The Case of

Connected Mathematics

Glenda Lappan
Elizabeth Difanis Phillips
James T. Fey

THE Connected Mathematics Project (CMP) was funded by the National Science Foundation between 1991 and 1997 to develop a mathematics curriculum for grades 6, 7, and 8.[1] The result was Connected Mathematics 1 (CMP 1), a complete mathematics curriculum that helps students develop understanding of important concepts, skills, procedures, and ways of thinking and reasoning in number, geometry, measurement, algebra, probability, and statistics. In 2000, the National Science Foundation funded a revision of the CMP materials, CMP 2, to take advantage of what we learned in the six years that CMP 1 had been used in schools. The following discussion is relevant to both CMP 1 and 2.

Connected Mathematics at a Glance

Below are some important features of the CMP curriculum:

- *It is organized around important mathematical ideas and processes.* The mathematics in the curriculum is carefully selected and sequenced to develop a coherent, connected curriculum.

The Connected Mathematics Project curriculum was developed by faculty at Michigan State University, the University of Maryland, and the University of North Carolina and is published by Prentice Hall. The co-principal investigators for the first edition were Glenda Lappan, James T. Fey, Susan N. Friel, William M. Fitzgerald, and Elizabeth A. Phillips. Codirectors for the second edition, CMP 2, are Glenda Lappan, James T. Fey, Susan N. Friel, and Elizabeth A. Phillips.

1. The work on which this chapter is based was funded in part by the National Science Foundation (NSF) through Grant No. ESI-9986372 to Michigan State University. Any opinions, findings, conclusions, or recommendations expressed here are those of the authors and do not necessarily reflect the views of the NSF.

- *It is problem-centered.* Important mathematical concepts are embedded in interesting problems to promote deeper engagement and learning for students. Students develop deep understanding of essential mathematical ideas, related skills, and ways of reasoning as they explore the problems individually, in a group, or with the class.

- *It connects mathematical ideas within a unit, across units, and across grade levels.* The development of new concepts is built on, and connected to, prior knowledge. Instead of seeing mathematics as a series of unrelated experiences, students learn to recognize how ideas are connected and develop a disposition to look for connections in the mathematics they study.

- *It provides practice with concepts and related skills.* The in-class development problems and the homework problems give students practice with important concepts, related skills, and algorithms, distributed over time.

- *It is for teachers as well as students.* The materials were written to support teachers' learning of both mathematical content and pedagogical strategies. The teachers' guides include extensive help with mathematics, pedagogy, and assessment. Multidimensional tasks are furnished in the assessment materials.

- *It is research-based.* Each unit was field-tested, evaluated, and revised over a six-year period. Approximately 200 teachers and 45,000 students in diverse school settings across the United States participated in the development of the curriculum.

The Influence of Theory and Research on the Development of Connected Mathematics

The curriculum, teacher support, and assessment materials that comprise the CMP program reflect influence from a variety of sources:

- knowledge of theory and research;
- authors' vision, imagination, and personal teaching and learning experiences;
- advice from teachers, mathematicians, teacher educators, curriculum developers, and mathematics education researchers;
- advice from teachers and students who used pilot and field-test versions of the materials.

The fundamental features of the CMP program—focus on big ideas of middle grades mathematics, teaching through student-centered exploration of mathematically rich problems, and continual assessment to guide instruction—reflect the distillation of advice and experience from those varied sources.

Our work was influenced in significant ways by what we knew of existing theory and research in mathematics education. Here we mention and explain briefly the essential themes in the theory and research basis for our work.

Research on Learning

Conceptual and procedural knowledge

We have been influenced by theory and research indicating that mathematical understanding is fundamentally a web of logical and psychological connections among ideas. Furthermore, we have interpreted research on the interplay of conceptual and procedural knowledge to say that sound conceptual understanding is an important foundation for procedural skill, not an incidental and delayed consequence of repeated, rote procedural practice.

Multiple representations

An important indication of students' connected mathematical knowledge is their ability to represent ideas in a variety of ways. We have interpreted this theory to imply that curriculum materials should frequently provide, and ask for, knowledge and information represented using graphs, number patterns, written explanations, and symbolic expressions.

Cooperative learning

A consistent and substantial body of research indicates that when students engage in cooperative work on appropriate problem-solving tasks, their mathematical and social learning will be enhanced. For example, a variety of conjectures and strategies can emerge from cooperative work as students work together to evaluate and refine their ideas. We have interpreted this line of theory and research to imply that we should design materials—for students and for teachers—that are suitable for use in cooperative learning instructional formats as well as individual learning formats. The mathematical task dictates the format.

Mathematics Education Research

Rational numbers and proportional reasoning

The extensive psychological literature on the development of rational numbers and proportional reasoning has guided our development of curriculum materials addressing these important middle school topics. Important understandings, such as the different interpretations of fractions, the role of the unit or whole, and partitioning and repartitioning a whole, have been built into the rational number strand. Furthermore, the implementation of CMP materials in real classrooms has allowed us to contribute to that literature with research publications that show the effects of new curriculum materials and teaching approaches to traditionally difficult topics.

Probability and statistical reasoning

The interesting research literature concerning the development of, and cognitive obstacles to, students' learning of statistical concepts like *mean* and *graphic displays* and probability concepts like the *law of large numbers* and *conditional probability* has been used as we developed the statistics and probability units of CMP materials.

Algebraic reasoning

The different conceptualizations of algebra described and researched in the literature contributed to the treatment of algebra in CMP. Different scholars describe algebra as a study of modeling, functions, and generalized arithmetic, and as a problem-solving tool. CMP has aspects of each of these descriptions of algebra but focuses more directly on functions and on the effects of rates of change among variables and on their representations. The research literature illuminates some of the cognitive complexities inherent in algebraic reasoning and offers suggestions on helping students overcome difficulties. Research concerning concepts such as *equivalence, functions, the equal sign, algebraic variables, graphical representations, multiple representations,* and *the role of technology* were used as we developed the algebra units of the CMP materials.

Geometric and measurement reasoning

Results from national assessments and research findings show that students' achievement in geometry and measurement is weak. Research on students' understanding and learning of geometric and measurement concepts such as *angle, area, perimeter, volume,* and processes such as *visualization* contributed to the development of the geometry and measurement units in CMP materials. As a result of research shifting from a focus on shape and form to the related ideas of congruence, similarity, and symmetry transformations, CMP geometry units were designed to focus on these important ideas.

Research from Education Policy and Organization

Motivation

One of the fundamental challenges in mathematics teaching is convincing students that serious effort in study of the subject will be rewarding and that the learning of mathematics can also be an enjoyable experience. We have paid careful attention to literature on extrinsic and intrinsic motivation, and we have done some informal developmental research of our own to discover aspects of mathematics and teaching that are most effective in engaging students' attention and interest.

Teacher and school change

The most attractive school mathematics curriculum materials will be of little long-term value or effect if there is not sufficient help for teachers in implementing the curriculum. In the process of helping teachers through professional development, we have paid close attention to what is known about teachers' effective professional development and the school implementation strategies that seem to be most effective.

Although each of these nine points indicates the influence of theory and research on design and development of the CMP curriculum, teacher, and assessment materials, it would be misleading to suggest that the influence is direct and controlling in all decisions. As the authors have read the research literature reporting empirical and theoretical work, research findings and new ideas have been absorbed and factored into the creative, deliberative, and experimental process that leads to a comprehensive mathematics program for schools.

Connected Mathematics:
A Curriculum for Students and Teachers

The CMP materials reflect the understanding that teaching and learning are not distinct—"what to teach" and "how to teach it" are inextricably linked. The circumstances in which students learn affect what is learned. The needs of both students and teachers are considered in the development of the CMP curriculum materials. This curriculum helps teachers and those who work to support teachers examine their expectations for students and analyze the extent to which classroom mathematics tasks and teaching practices align with their goals and expectations.

Overarching Goal of Connected Mathematics

"The overarching goal of CMP is to help students and teachers develop mathematical knowledge, understanding, and skill along with an awareness of and appreciation for the rich connections among mathematical strands and between mathematics and other disciplines" (Lappan 2006b, p. 2). The single mathematical standard that has been a guide for all the CMP curriculum development is the following:

> All students should be able to reason and communicate proficiently in mathematics. They should have knowledge of and skill in the use of the vocabulary, forms of representation, materials, tools, techniques, and intellectual methods of the discipline of mathematics, including the ability to define and solve problems with reason, insight, inventiveness, and technical proficiency (Lappan 2006b, p. 2).

Identifying Exit Goals

We began our development work with some assumptions about what students would know when they entered grade 6. Then, working with our advisory board—mathematicians, scientists, mathematics educators, teachers, parents, and business representatives—our next task was to define the knowledge and skills in each content strand that students should have by the time they leave grade 8. To aid in the process and help us think more broadly about the curriculum, we wrote a set of papers that outlined the exit goals for grade 8 in the four mathematical strands that we would develop—number, algebra, probability and statistics, and geometry and measurement. These papers were our touchstone for the development of each strand.

Principles That Guided the Development of Connected Mathematics

The authors were guided by the following principles in the development of the CMP materials. These statements reflect both research and policy stances in mathematics education about what works to support students' learning of important mathematics to levels higher than those we have achieved in the United States in the past.

- An effective curriculum has coherence: it builds and connects from problem to problem, investigation to investigation, unit to unit, and grade to grade. Thus, the essential mathematical ideas around which the curriculum is built are identified. Then the underlying concepts and related skills and procedures supporting the development of an essential idea are identified and included in an appropriate development sequence of problems.
- Classroom instruction focuses on inquiry and the investigation of mathematical ideas embedded in rich problem situations.
- Mathematical tasks for students in class and in homework are the primary vehicle for students' engagement with the mathematical concepts to be learned. The essential mathematical goals are elaborated, exemplified, and connected through the problems in an investigation.
- The curriculum develops both conceptual and procedural knowledge with the underlying assumption that the interaction of conceptual and procedural knowledge is what produces fluency.
- Ideas are explored through these tasks in the depth necessary to allow students to make sense of them. Superficial treatment of an idea produces shallow and short-lived understanding and does not support making connections among ideas.
- The curriculum helps students grow in their ability to reason effectively with information represented in graphic, numeric, symbolic, and verbal forms and to move flexibly among these representations.
- The curriculum reflects the information-processing capabilities of calculators and computers and the fundamental changes such tools are making in the way people learn mathematics and apply their knowledge of problem-solving tasks.

CMP is different from more conventional curricula in that it is problem-centered. The following section elaborates what we mean by this and what the added value is for students of such a curriculum.

Rationale for a Problem-Centered Curriculum

Students' perceptions about a discipline come from the tasks or problems with which they are asked to engage. For example, if students in a geometry course are asked to memorize definitions, they think geometry is about memorizing definitions. If students spend a majority of their mathematics time practicing paper-and-pencil computations, they come to believe that mathematics is about calculating answers to arithmetic exercises as quickly as possible. They may become faster at performing specific types of computations, but they may not be able to apply these skills to other situations or to recognize problems that call for these skills.

Formal mathematics begins with undefined terms, axioms, and definitions and deduces important conclusions logically from those starting points. However, mathematics is produced and used in a much more complex combination of exploration, experience-based intuition, and reflection. If the purpose of studying mathematics is to be able to solve a variety of problems, then students need to spend significant portions of their mathematics time solving problems that require thinking, planning, reasoning, computing, and evaluating.

Evidence from cognitive sciences' research supports the theory that students can make sense of mathematics if the concepts and skills are embedded in a context or problem. If time is spent exploring interesting mathematics situations, reflecting on solution methods, examining why the methods work, comparing methods, and relating methods to those used in previous situations, then students are likely to build more robust understanding of mathematical concepts and related procedures. This method is quite different from the assumption that students learn by observing a teacher as he or she demonstrates how to solve a problem and then practicing that method on similar problems.

A problem-centered curriculum not only helps students make sense of the mathematics, it also helps them process the mathematics in a retrievable way. Teachers who use CMP report that students in succeeding grades remember and refer to a concept, technique, or problem-solving strategy by the name of the problem in which they encountered the ideas. For example, the basketball problem from "What Do You Expect?" in the grade 7 curriculum becomes a trigger for remembering the processes of finding compound probabilities and expected values.

Results from research in the cognitive sciences also suggest that learning is enhanced if it is *connected* to prior knowledge and is more likely to be retained and applied to future learning. Critically examining, refining, and extending conjectures and strategies are also important aspects of becoming *reflective* learners.

In CMP, important mathematical ideas are embedded in the context of interesting problems. As students explore a series of connected problems, they develop understanding of the embedded ideas and, with the aid of the teacher, abstract powerful mathematical ideas, problem-solving strategies, and ways of thinking. They learn mathematics, and they learn *how* to learn mathematics.

Characteristics of Good Problems

To be effective, problems must embody crucial concepts and skills and have the potential to engage students in making sense of mathematics. And, since students build understanding by reflecting, connecting, and communicating, the problems need to encourage them to use these processes.

Each problem in CMP curricula satisfies the following criteria:

- The problem must have important, useful mathematics embedded in it.
- Investigating the problem should contribute to students' conceptual development of important mathematical ideas.
- Work on the problem should promote skillful use of mathematics and opportunities to practice important skills.
- The problem should create opportunities for teachers to assess what students are learning and where they are experiencing difficulty.

In addition each problem satisfies some of or all the following criteria:

- The problem should engage students and encourage classroom discourse.
- The problem should allow a variety of solution strategies or lead to alternative decisions that can be taken and defended.
- Solving the problem should require higher-level thinking and problem solving.
- The mathematical content of the problem should connect with other important mathematical ideas.

Fluency with Concepts and Related Skills and Algorithms in Connected Mathematics

Students need to practice using new mathematics concepts, ideas, and procedures to reach a level of fluency that allows them to "think" with the ideas in new situations. To accomplish this we were guided by the following principles related to skills practice.

- Immediate practice should be related to the situations in which the ideas have been developed and learned.
- Continued practice should use skills and procedures in situations that connect with ideas that students have already encountered.
- Students need opportunities to use the ideas and skills in situations that extend beyond familiar situations. They need to recognize uses of skills and algorithms in the development of additional mathematics. These opportunities allow students to use skills and concepts in new combinations to develop new concepts and procedures and solve new kinds of problems.
- Students need practice distributed over time to allow ideas, concepts, and procedures to reach a level of fluency of use in familiar and unfamiliar situations and to allow connections to be made with other concepts and procedures.
- Students need guidance in reflecting on what they are learning, how the ideas fit together, and how to make judgments about what is helpful in which kinds of situations.
- Students need to learn how to make judgments about what operation or combination of operations or representations is useful in a given situation, as well as become skillful at carrying out the needed computation(s). Knowing *how* to, but not *when* to, is insufficient.

Rationale for Depth versus Spiraling

The concept of a "spiraling" curriculum is philosophically appealing, but, too often, not enough time is spent initially with a new concept to build on it at the next stage of the spiral. This leads to teachers' spending a great deal of time reteaching the same ideas over and over again. Without a deeper understanding of concepts and how they are connected, students come to view mathematics as many different techniques and algorithms to be memorized. Problem solving based on such learning becomes a search for the correct algorithm instead of seeking to make sense of the situation, considering the nature and size of a solution, putting together a solution path that makes sense, and examining the solution in light of the original question. Taking time to allow the ideas studied to be more carefully developed means that when these ideas are met in future units, students have a solid foundation on which to build. Instead of being caught in an endless cycle of relearning the same ideas to some superficial, quickly forgotten level, students are able to connect new ideas with previously learned ideas and make substantive advances in the depth of understanding and use of these ideas.

With any important mathematical concept, there are many related ideas, procedures, and skills. In CMP at a grade level, students study a small, select set of important mathematical concepts, ideas, and related procedures in depth rather than skim through a larger set of ideas in a shallow manner. This means that time is allocated to developing the understanding of key ideas in contrast with "covering" a book.

Developing a Depth of Understanding and Use

Through the field trials process we were able to develop units that result in student understanding of key ideas in depth. An example is illustrated in the way that CMP treats proportional reasoning—a fundamentally important topic for middle school mathematics and beyond. Conventional treatments of this central topic are often limited to a brief expository presentation of the ideas of ratio and proportion, followed by training in techniques for solving proportions. In contrast, the CMP curriculum materials develop core elements of proportional reasoning in a seventh-grade unit, "Comparing and Scaling," with the groundwork for this unit being developed in four prior units. Six succeeding units build on and connect with students' understanding of proportional reasoning.

Teachers' Guides and Support Materials

When mathematical ideas are embedded in problem-based investigations of rich contexts, the teacher has a crucial responsibility for ensuring that students abstract and generalize the important mathematical concepts and procedures from the experiences with the problems. In a problem-centered classroom, teachers take on new roles—moving from always being the one who does the mathematics to being the one who guides, questions, and facilitates the learner in doing and making sense of the mathematics.

The teachers' support materials engage teachers in a conversation about what is possible in the classroom around a particular lesson.

- The goals of each lesson are articulated, and suggestions are made about how to engage the students in the mathematics task, how to promote students' thinking and reasoning during the exploration of the problem, and how to summarize with the students the important mathematics embedded in the problem.

- An overview and elaboration of the mathematics of the unit is included along with examples and a rationale for the models and procedures used. This mathematical essay is to help a teacher stand above the unit and see the mathematics from a perspective that includes the particular unit. The essay also connects with earlier units, and it projects to where the mathematics goes in subsequent units and years.

- Actual classroom scenarios are included to help stimulate teachers' imaginations about what is possible.

- Questions to ask students at all stages of the lesson are included to help teachers support students' learning.

- Suggestions are provided for involving students with special needs and those who are English language learners.

- Strategies for supporting students' learning through group work are included.

- Reflection questions are given at the end of an investigation to help teachers assess what sense students are making of the big ideas and to help students abstract, generalize, and record the mathematical ideas and techniques developed in the investigation.

- Multiple kinds of assessment and grading help are included to help teachers see assessment and evaluation as a way to inform students of their progress and parents of students' progress, but also to guide the decisions a teacher makes about lesson plans and classroom interactions. More demanding partner quizzes are used to mirror classroom practices as well as highlight important concepts, skills, techniques, and problem-solving strategies.

- A Web site for parents furnishes information on individual units and the lesson problems in a unit. Aid is given to support parents in helping students with homework.

Developing CMP 2: Repeating the Cycle of Research, Field Testing, and Evaluation

Before starting the design phase of the second-edition materials, we commissioned reviews of our earlier CMP 1 units from eighty-four individuals in seventeen states and comprehensive reviews from more than twenty schools in fourteen states. The individual reviews focused on particular strands over all three grades or on particular subpopulations such as special-needs or underserved students, or on overall concerns such as language use and readability. The other reviews were conducted by groups of teachers who came together to review a grade level of the curriculum. The people involved in this collection of reviews were teachers, administrators, curriculum supervisors, mathematicians, and experts in special education, language and reading level analyses, English language learners, issues of equity, and others. All their responses were coded and entered into a database that allowed reports on feedback to be printed for any issue or combination of issues that would be helpful to an author or staff person in designing a unit.

Additionally, CMP issued a call to schools to serve as pilot schools for the development of CMP 2. We received fifty applications from districts for piloting. From these applications we picked fifteen that included forty-nine school sites in twelve states and the District of Columbia. We received evaluation feedback from these sites over the five-year cycle of development.

Based on the commissioned reviews, what the authors had learned from schools that used CMP over a six-year period, and our advisory board, the authors started with grades 6 and 7 and systematically revised and restructured the units and their sequence for each grade level to create a first draft of the revision. These were sent to our pilot schools to be taught during the second year of the project. These initial grade-level unit drafts were the basis for substantial feedback from our trial teachers.

The units for each grade level went through at least three cycles of field trials to feedback to revision to field trials. When needed, units went through an extra iteration of field trials. This process of (1) commissioning reviews from experts, (2) using the field trials–feedback loops for the materials, (3) conducting essential classroom observations of units being taught, and (4) monitoring students' performance on state and local tests by trial schools, comprises research-based development of curriculum. This process takes five years to produce the final drafts of units that are sent to the publisher. Another eighteen months is needed for editing, design, and layout for the published units. This process produces materials that are cohesive and effectively sequenced. Figure 5.1 below shows the cycle we used for both CMP 1 and CMP 2.

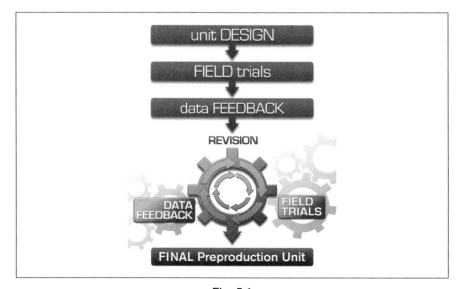

Fig. 5.1

Connected Mathematics:
A Curriculum Codeveloped with Teachers and Students

Developing a curriculum with a complex set of interrelated goals takes time and input from many people. Our work was based on a set of deep commitments that we, as authors, had about what would constitute a more powerful way to engage students in making sense of mathematics. Our advisory board took an active role in reading and critiquing units in their various iterations. In order to enact our development principles, we found that three full years of field trials in schools were essential. This feedback from teachers and students was the pivotal element in the success of the CMP materials. The final materials comprised the ideas that stood the test of time in classrooms. Nearly 200 teachers in fifteen trial sites around the country, and their thousands of students, were a significant part of the team of professionals that made these materials happen. The scenarios of teachers' and students' interactions with the materials became the most compelling parts of the Teachers' Guides.

From the authors' perspectives, our hope was to develop materials that play out deeply held beliefs and firmly grounded theories about what mathematics is important for students to learn and how they should learn it. We hope that we have been a part of helping to challenge and change the curriculum landscape of our country.

BIBLIOGRAPHY

Ben-Chaim, David, James T. Fey, William M. Fitzgerald, Catherine Benedetto, and Jane Miller. "Proportional Reasoning among 7th Grade Students with Different Curricula Experiences." *Educational Studies in Mathematics* 36 (September 1998): 247–73.

Kilpatrick, Jeremy, W. Gary Martin, and Deborah Schifter, eds. *A Research Companion to "Principles and Standards for School Mathematics."* Reston, Va.: National Council of Teachers of Mathematics, 2003.

Lambdin, Diana V., and Jane M. Keiser. "The Clock Is Ticking: Time Constraint Issues in Mathematics Teaching Reform." *Journal of Educational Research* 90 (September–October 1996): 23–32.

Lambdin, Diana V., and Ronald V. Preston. "Caricatures in Innovation: Teacher Adaptation to an Investigation-Oriented Middle School Mathematics Curriculum." *Journal of Teacher Education* 46 (March 1995): 130–40.

Lamon, Susan J. *Teaching Fractions and Ratios for Understanding.* Mahwah, N.J.: Lawrence Erlbaum Associates, 1999.

Lappan, Glenda, William M. Fitzgerald, Elizabeth D. Phillips, Janet P. Schroyer, and Mary Jean Winter. *The Middle Grades Mathematics Project.* Menlo Park, Calif.: Addison-Wesley, 1986.

Lappan, Glenda, James T. Fey, Susan N. Friel, William M. Fitzgerald, and Elizabeth D. Phillips. *Connected Mathematics.* Needham, Mass.: Prentice Hall, 2002.

———. *Connected Mathematics 2.* Boston, Mass.: Prentice Hall, 2006a.

———. *Implementing and Teaching Guide.* Boston, Mass.: Prentice Hall, 2006b.

National Council of Teachers of Mathematics (NCTM). *Curriculum and Evaluation Standards for School Mathematics.* Reston, Va.: NCTM, 1989.

————. *Professional Standards for Teaching Mathematics.* Reston, Va.: NCTM, 1991.

————. *Assessment Standards for School Mathematics.* Reston, Va.: NCTM, 1995.

————. *Principles and Standards for School Mathematics.* Reston, Va.: NCTM, 2000.

Ridgway, James E., Judith S. Zawojewski, Mark N. Hoover, and Diana V. Lambdin. "Student Attainment in the *Connected Mathematics* Curriculum." In *Standards-Based School Mathematics Curricula: What Are They? What Do Students Learn?* edited by Sharon L. Senk and Denisse R. Thompson, pp. 193–224. Mahwah, N.J.: Lawrence Erlbaum Associates, 2003.

Streefland, Leen. *Fractions in Realistic Mathematics Education.* Dordrecht, Netherlands: D. Reidel, 1991.

6

The Case of

Mathematics in Context

David C. Webb
Margaret R. Meyer

THE Mathematics in Context (MiC) curriculum project was funded in 1991 by the National Science Foundation to develop a comprehensive mathematics curriculum for grades 5–8. Collaborating on this project were research and development teams from the National Center for Research in Mathematical Sciences Education at the University of Wisconsin—Madison and the Freudenthal Institute (FI) at the University of Utrecht in the Netherlands. This collaboration was initiated by Thomas Romberg, who had become familiar with the work of the Freudenthal Institute while chairing the writing team of the NCTM's *Curriculum and Evaluation Standards for School Mathematics* (1989). He recognized that the mathematics curricula used in the Netherlands was consistent with the vision of the emerging *Standards* and that it could serve as a model for a middle grades curriculum for the United States.

Design Principles

The design principles for MiC are derived from Realistic Mathematics Education (RME). Since 1971, researchers at FI have developed and applied this theoretical approach toward the learning and teaching of mathematics. RME incorporates views on what mathematics is, how students learn mathematics, and how mathematics should be taught. The principles that underlie this approach are strongly influenced by Hans Freudenthal's concept of "mathematics as a human activity" (Freudenthal 1983). From the RME perspective, students are

This chapter is a synthesis of several reports and papers authored by Thomas Romberg of the University of Wisconsin—Madison and Jan de Lange of the Freudenthal Institute at the University of Utrecht, Netherlands, co-principal investigators for the Mathematics in Context development project.

seen as reinventors, with teachers guiding and making conscious to students the mathematization of reality, with an eye to encouraging students to reflect on the process. The process of guided reinvention is supported by students' engagement in problem solving, a collective as well as an individual activity, in which whole-class discussions centering on conjecture, explanation, and justification play a crucial role. The instructional guidance of teachers in this process is essential, because teachers gradually introduce and negotiate with the students the meanings and use of conventional mathematical terms, symbols, representations, and procedures.

RME's set of philosophic tenets transfers into a design approach for MiC with the following components: (1) developing instruction based in experientially real contexts; (2) designing structured sets of instructional activities that reflect and work toward important mathematical goals; (3) designing opportunities to build connections between content strands, through solving problems that reflect these interconnections; (4) building on students' prior conceptions and recognizing and making use of their representations (models) to support the development of more formal mathematics (i.e., progressive formalization); (5) designing activities to promote pedagogical strategies that support students' collective investigation of reality; (6) posing questions and prompts that evoke various levels of students' mathematical reasoning; and (7) developing an assessment system that monitors the progress of both the group and individual students.

Starting in Reality

The starting point of any instructional sequence should involve situations that are experientially real to students so that they can immediately engage in personally meaningful mathematical activity (Gravemeijer 1990). Such problems often involve everyday life settings or fictitious scenarios, although mathematics itself can also serve as a context of interest. Such activities should either reflect real phenomena from which mathematics has developed historically or actual situations and phenomena in which further interpretation, study, and analysis requires the use of mathematics.

Identifying the Learning Lines

A second tenet of RME is that the starting point should also be justifiable by the potential end point of a learning sequence. To accomplish this, the mathematical domain needs to be well mapped. This involves identifying the important features and resources of the domain that are important for students to find, discover, use, or even invent for themselves, and then relating them through long learning lines (van den Heuvel-Panhuizen 2002). This is not a trivial task, as Webb and Romberg (1992, p. 47) argued:

> Knowledge of a domain is viewed as forming a network of multiple possible paths and not partitioned into discrete segments.... Over time, the maturation of a student's functioning within a domain should be observed by noting the formation of new linkages, the variation in the situations the person is able to work with, the degree of abstraction that is applied, and the level of reasoning applied.

The situations that serve as starting points for a domain are crucial and should continue to function as paradigm cases that involve rich imagery and, thus, anchor students' increasingly abstract activity (Meyer, Dekker, and Querelle 2001). The students' initially informal mathematical activity should constitute a basis from which they can abstract and construct increasingly sophisticated mathematical conceptions.

Interconnectivity

The third tenet of RME is based on the observation that real phenomena in which mathematical structures and concepts manifest themselves lead to interconnections within and between content strands as well as connections with other disciplines (e.g., biological sciences, physics, sociology, and so on). Although the maps developed for each of the four MiC content strands—number, algebra, geometry and measurement, and data analysis—contain unique terms, representations, and procedures, instruction in actual classrooms inevitably involves the intertwining of these strands. Problems can often be viewed from multiple viewpoints. For example, a geometric pattern can be expressed using number relationships or algebraically.

Students' Creation of Mathematical Models

RME's fourth tenet is that instructional sequences should involve activities in which students reveal and create models of their informal mathematical activity. RME's heuristic for laying out long learning lines for students involves a conjecture about the role that emergent models play in the students' learning, namely that students' models of their informal mathematical activity can evolve into models for increasingly abstract mathematical reasoning (Gravemeijer 1991). Gravemeijer (1994, p. 102) explained this bottom-up progression with four levels of progressive mathematization (see fig. 6.1).

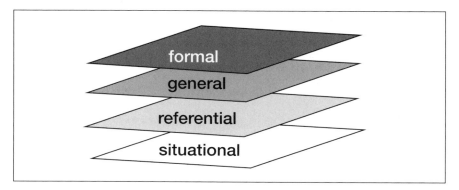

formal

general

referential

situational

Fig. 6.1. Four levels of mathematization

At the initial, "situational" level, the expectation is that students develop interpretations, representations, and strategies appropriate for engaging with a particular problem context. Paul Cobb (1994, pp. 23–24), in his analysis of the RME approach to instruction, states:

> Instructional sequences should involve activities in which students create and elaborate symbolic models of their informal mathematical activity. This modeling activity might involve making drawings, diagrams, or tables, or it could involve developing informal notations or using conventional mathematical notations.

At the "referential" level, students create informal *models of* the problem situation. Such *models-of* contain the collective descriptions, concepts, procedures, and strategies that refer to concrete or paradigmatic situations. At the "general" level, as a result of generalization, exploration, and reflection, students are expected to mathematize their informal modeling activity and begin to focus on interpretations and solutions independent of situation-specific imagery. Models at this level are considered *models-for* and are used as a basis *for* reasoning and reflection. The "formal" level involves reasoning with conventional symbols and is no longer dependent on *models-for*.

Thus, for students, a model is first constituted, for example, as a context-specific model of a situation, and then generalized across situations. In this process, the model changes in character and becomes an entity in itself, functioning eventually as a basis for mathematical reasoning on a formal level (Meyer 2001). The development of ways of symbolizing problem situations and the transition from informal to formal notations are important aspects of the selection of problem contexts, the relationships between contexts, and instructional assumptions.

The consequences are that the introduction of numbers, number sentences, standard algorithms, terms, signs and symbols, and the rules of use in formal mathematics should involve a process of social negotiation similar to how the notations and rules were originally developed.

The RME instructional approach has been described as "bottom-up," in that students construct models for themselves that then serve as the basis for developing formal mathematical knowledge. A significant challenge, therefore, in the design process is to find, adapt, or create a collection of problem situations and organize them in a structured manner that both engages students, so that they explore each domain in this manner, and allows for growth in what is to be learned. Furthermore, the design of these problem contexts and their sequencing requires observation of how children engage in the proposed problem contexts in real classrooms. Romberg (1992, p. 778), for example, argued:

> Too often a problem is judged to be relevant through the eyes of adults, not students. Also, this perception is undoubtedly a Western, middle-class, static vision. Concrete situations, by themselves, do not guarantee that students will see relevance to their worlds; they may not be relevant for all students nor prepare them to deal with a changing, dynamic world.

Relevance is not the sole criteria for selecting problems. The relation between context and mathematics also poses major challenges (de Lange 1987). Although there is no doubt that many interesting activities exist or can be created, not all lead to growth of students' knowledge of the mathematical concepts involved. The challenge is to design a sequence of activities such that students grow in their knowledge and understanding of the ideas in a domain over time. Each activity, therefore, has to be justifiable by potential end points in a learning sequence.

This design principle of sequencing activities and questions to facilitate students' development of more formal models, representations, and strategies is the essence of progressive formalization. The National Research Council report *How People Learn* (Bransford, Brown, and Cocking 1999) describes progressive formalization as one of the "approaches to the development of curricula that supports learning with understanding and encourages sense making" (p. 125). Progressive formalization describes a learning sequence that begins with informal strategies and knowledge, developing these into preformal methods that are still linked with concrete experiences, models, and strategies. Then, through the process of guided reinvention (guided by the instructional materials and the teacher or facilitator), the preformal models and strategies progressively develop into more formal and abstract mathematical procedures, concepts, and insights.

As an example of what these informal and preformal representations entail, the following is an abbreviated sequence of algebraic representations from the MiC unit "Building Formulas" (Wijers et al. 2006). This activity focuses on relationships among geometric and numerical patterns given in tabular form, related recursive and direct formulas, and equivalent representations of direct formulas. This activity begins with exploration of the relationship between the length of a beam and the number of rods used to make the beam.

The sequence of beams is organized into a table (fig. 6.2), and teachers are asked to discuss any patterns they might see. This leads teachers to revisit the definition of a recursive formula, identified elsewhere in MiC as a NEXT-CURRENT formula. A recursive formula is a preformal representation in which the NEXT term for a sequence is found by operating on the CURRENT term. In the instance of the beams context, the number of rods in the NEXT beam is found by adding 4 to the number of rods in the CURRENT beam (i.e., NEXT = CURRENT + 4). Although recursive formulas are useful for discussing numerical patterns, they are not very helpful when trying to find the number of rods for very large beams. This makes it necessary to develop a direct formula ($R = 4L - 1$ or other equivalent formulas).

	Length of Beam (L)	Number of Rods (R)
	1	3
	2	
	3	
	4	

Fig. 6.2. Table of beams and rods

The repeated additive operation of $+4$ in the recursive formula for this context is represented by the operation of $\times 4$ in the direct formula (repeated addition is once again related to multiplication, in an algebraic context). The relationship between the increments in the table and the additive operation in the recursive formula, and their relationship to slope, are explored further in a different activity involving graphs of linear phenomena. In this way, informal and preformal representations provide a well-connected foundation for understanding formal algebraic representations—that is, progressive formalization.

Later in the same activity (fig. 6.3), pictures of the beams are used to discuss and justify the equivalence of various direct formulas. These informal and preformal representations are more than mathematical stepping stones; in MiC they serve as valuable referents used to make sense of formal representations.

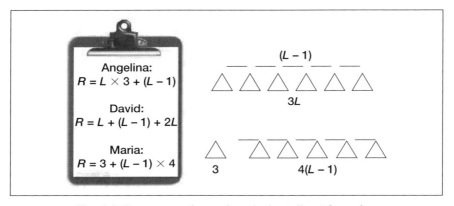

Fig. 6.3. Representations of equivalent direct formulas

Interactive Instruction

RME's fifth tenet is that in classrooms the learning process can be maximized when it occurs within the context of interactive instruction. Students are expected to explain and justify their solutions, to work to understand other students' solutions, to agree and disagree openly, to question alternatives, and to reflect on what they have discussed and learned. Creating a classroom that fosters such interaction involves a shift in the teacher's role—from dictating the prescribed knowledge through a routine instructional sequence to orchestrating activities situated in learning trajectories wherein ideas emerge and develop in the social setting of the classroom. To promote the interaction of students' representations and strategies, teachers must also create discourse communities that support and encourage students' conjectures, modeling, remodeling, and argumentation (e.g., Cobb et al. 1997).

The teacher's role in the RME instructional process involves capitalizing on students' reasoning and continually introducing and negotiating with students the emergence of shared terms, symbols, rules, and strategies, with an eye to encouraging students to reflect on what they learn (Ball 1993). Students are seen as reinventors, with the teacher guiding and making conscious to students the mathematization of reality, with symbolizations emerging and developing meaning in the social situation of the classroom. Students are encouraged to communicate their knowledge, either verbally, in writing, or through other means like pictures, diagrams, or models to other students and the teacher. Central to this interactive classroom is the development of students' abilities to use mathematical argumentation to support their own conjectures.

Levels of Reasoning

The responsibility for promoting students' understanding cannot rest entirely on the teacher's ability to respond to "teachable moments" as students engage in rich mathematical activities. Curriculum-embedded prompts and questions can also facilitate the development of students' understanding. Throughout MiC, different levels of questions, requiring different cognitive responses from students, are posed to elicit sustained reasoning in order to promote deeper understanding of the content. Questions that elicit different levels (or types) of reasoning are not necessarily more difficult. Rather, they require a different type of thinking. The different levels of questions posed throughout MiC can be characterized as follows.

Level 1: Reproduction, procedures, concepts, and definition

This level deals with knowing facts, representing, recognizing equivalents, recalling mathematical objects and properties, performing routine procedures, applying standard algorithms, and developing technical skills, as well as dealing and operating with statements and expressions containing symbols and formulas in "standard" form.

Level 2: Connections and integration for problem solving

At this level, students start making connections within and between the different domains in mathematics, integrate information in order to solve simple problems, and have a choice of strategies and a choice in the use of mathematical tools. At this level, students are also expected to handle different representations appropriate to the situation and purpose and need to be able to distinguish and relate a variety of statements (e.g., definitions, examples, conditional assertions, proofs).

Level 3: Mathematization, mathematical thinking, generalization, and insight

At this level, students are asked to mathematize situations, recognize and extract the mathematics embedded in the situation and use mathematics to solve the problem, analyze, interpret, develop models and strategies, and make mathematical arguments, proofs, and generalizations.

Assessment in MiC

It is worth emphasizing that the content and reasoning goals for MiC instruction are consistent with its assessment program. Students are given opportunities to demonstrate these higher reasoning levels as they mathematize problem contexts, solve new problems, and justify strategies throughout each unit as well as on mid-unit or end-of-unit assessments.

The RME approach to formative assessment is closely aligned with instruction and is seen as part of daily instructional practice (de Lange 1987; van den Heuvel-Panhuizen 1996). Ascertaining how many artifacts of the domain a student can identify is insufficient evidence of the type of students' knowledge and understanding that can be used to guide instruction. Rather, assessment should focus on the ways students identify and use such artifacts to model, solve, and defend their solutions with respect to increasingly complex tasks. Assessment tasks need to reveal to teachers the representations, strategies, and assumptions students make when solving problems. Monitoring students' progress, therefore, involves the use of open tasks, through which students relate concepts and procedures and use them to solve nonroutine problems. This is in contrast with tasks that require the reiteration of procedures "learned" to solve problems that merely mimic the content covered. Responses to open tasks should provide reliable evidence of what a student is able to do in any domain at a given point in the instructional sequence.

In 1999, de Lange and the FI staff developed a *Framework for Classroom Assessment in Mathematics,* which both reflect RME's approach to assessment and is consistent with assessment principles outlined in NCTM's *Assessment Standards* (1995). The intent of this framework was to generalize what had been learned about classroom assessment into a set of principles, with examples, so that the ideas could be used by teachers at all levels using any curricular materials.

The following are principles of the framework.

1. The main purpose of classroom assessment is to improve learning.
2. The mathematics is embedded in worthwhile (engaging, educative, authentic) problems that are part of the students' real world.
3. Assessment methods should enable students to show what they know rather than what they do not know.
4. Assessments include multiple and varied opportunities (formats) for students to display and document their achievement.
5. Tasks make operational all the goals of a curriculum (not just the "lower" ones). Performance standards, including indications of the different levels of mathematical thinking, are helpful tools in this process.
6. Grading criteria, including examples of exemplary and less-than-exemplary work, are published and consistently applied.
7. Testing and grading involve minimal secrecy.
8. Feedback given to students is genuine.
9. The quality of a task is not defined by its accessibility to objective scoring, reliability, or validity in the traditional sense, but by its authenticity, fairness, and exemplification of the principles above.

Teachers who desire to teach for students' understanding recognize the need for a broader perspective of classroom assessment, as articulated by these principles. From a content perspective, it is also important to recognize that many skills are drawn on simultaneously when doing real mathematics. Students' experiences in school mathematics, therefore, should motivate them to draw on their full range of abilities. With respect to assessment, students should have an opportunity to demonstrate the extent to which they are able to apply their mathematical knowledge in new ways. The assessment principles for RME are based on the assumption that an overarching goal of any assessment program should be the support of teaching for students' understanding (see also Romberg [2004]).

The seamless nature of assessment and instruction in MiC accentuates the importance of conveying to teachers the design principles for assessment and how they are actualized throughout the program. The way in which teachers conceptualize assessing students' progress has a strong influence on their instructional decisions (Her and Webb 2004; Webb 2004). One of the more dramatic differences in teachers' guide content for the first and second editions of MiC was in the way the different types of reasoning expectations for questions were made more explicit, in terms of how these goals were addressed in ongoing assessment opportunities and unit assessments.

Mathematics in MiC

The tenets of RME guided not only how mathematics should be introduced, developed, and formalized but also to some extent the content of mathematics to be included. The interplay between desired mathematical endpoints and known learning lines resulted in a set of potential problem contexts, preformal models and strategies, and related instructional pathways. Although this interplay articulated a significant portion of the mathematical content in MiC, the cross-national perspectives represented among the development team influenced the response to fundamental psychological and epistemological questions such as the following: What is geometry? How can children use mathematics to make sense of their world? What is the reasoning that underlies the formal, symbolic representations and strategies associated with algebra?

Decisions regarding mathematical topics often reflected the mathematical reasoning students were capable of as guided by the developmental research of the Freudenthal Institute and as demonstrated in the field testing of MiC units. For example, several of the MiC algebra units in the restrictions substrand included problems and techniques associated with covariance and linear programming to develop students' understanding of linear equations, inequalities, and systems of equations. Even though the content of these units went beyond what is typically taught in middle grades mathematics, students regularly demonstrated sophisticated algebraic reasoning when completing these activities (Webb and Meyer 2002). The MiC development team was convinced that such topics should remain in the program to lay a stronger foundation for the study of secondary school mathematics.

Developing Mathematics in Context

MiC was initiated with the preparation of a blueprint document by an international advisory committee to guide the mathematical content addressed in the materials. The FI staff, under the direction of Jan de Lange, prepared initial drafts of individual units based on the blueprint. Researchers at the University of Wisconsin—Madison then modified the language and problem contexts in these units to make them appropriate for students and teachers in the United States. Pilot versions of the individual units were tested in middle schools in Wisconsin, and then field-test versions of the units were created from feedback received during the pilot. The field-test versions were then used in several states and Puerto Rico, and data from the field-test sites were used to help write revisions to students' books and teachers' guides before commercial publication.

During the pilot and field-test phases, teachers' use of and students' responses to each unit were monitored by author teams, which included at least one Dutch and one American coauthor. These extended classroom studies resulted in data collection that included many hours of videotaped lessons, field notes from observations, interviews with teachers, and students' work. Students' achievement data from both the pilot and field-test phases were also analyzed (cf. Webb and Meyer 2002). After evaluating several teachers' implementations of a particular unit, author teams proposed adjustments to the sequence of mathematical content and changes to problem contexts. Such modifications included how content was introduced, developed, and emphasized across several related units, within or across content strands. Results from many of these extended studies of teachers' use of MiC were later disseminated as dissertations, book chapters (e.g., Shafer and Romberg 1999; Webb et al. 2005), edited books (Romberg 2004), and journal articles (e.g., van Reeuwijk and Wijers 1997; Meyer 1999).

One of the most significant challenges during the pilot and field-testing of MiC was the need to prepare teachers whose experiences with, and beliefs about, teaching and learning mathematics were inconsistent with design principles of the curriculum. Even though many of the field-test teachers were viewed by their colleagues as innovators, when students engaged (and struggled) with new problem situations, teachers would follow established teaching routines and tend to "take over" to teach students how to solve the problem. Not surprisingly, teacher-centered instructional routines resulted in limited discourse by students and a reduction in the intended cognitive demand of problems, which also became problematic for assessment. Teachers often noted that the assessment tasks expected students to "take the mathematics to a new level" and were reluctant to include higher-level reasoning tasks on a quiz or unit test. Teachers were caught off guard with problems that had built-in thresholds for students' reasoning. This feature often raised the level of teachers' uncertainty and anxiety (e.g., Frykholm 2004), as they sought alternatives to the common practice of "taking students through a problem one step at a time."

Author teams worked directly with teachers to negotiate the tension between teachers' perceptions of what it meant to implement MiC successfully with their students and the developers' conceptions of how to achieve the mathematical goals and cognitive intent of a given unit. Since revisions depended, in part, on the author teams' evaluation of students' responses to the development of mathematical ideas, the interaction between teachers' use and students' response had to be taken into account when proposing necessary revisions. Adding a layer of complexity to the pedagogical challenges of implementation was the introduction of new mathematical models (e.g., ratio table, combination chart, notebook notation, arrow language, and so on). Teachers' discomfort with pedagogical change was compounded by their lack of familiarity with students' strategies that were elicited. To ameliorate these expected limitations in teachers' knowledge for effective pedagogical decisions, teachers were often reminded by authors to "trust the curriculum."

During the development of the first edition, many MiC coauthors were involved in professional development efforts at the field-test sites and, therefore, were sensitive to teachers' responses to the curriculum. Following the publication of the first edition in 1998, requests for program refinement were most often related to teachers' guide issues such as more detailed planning and pacing guides for each unit, more explicit identification of "crucial" problem contexts and mathematical models developed in each unit, and the need for additional assessment tasks. Several of the MiC authors, with support provided by Encyclopædia Britannica and the National Science Foundation–funded Show-Me Center, responded to various requests to develop, pilot-test, and distribute auxiliary materials. Some materials, such as "Core Concept Outlines" and "MiC Pacing Guides" were made available through the MiC Web site. Additional assessments—that is, *MiC Balanced Unit Assessments* (Dekker et al. 2000)—were published and made available for purchase. These materials and other "improvements by request" were later integrated into the teachers' guide content for the second edition of MiC.

From First to Second Edition: Changes to the Development and Revision Processes

An advisory committee that included many of the original coauthors prepared the blueprint for the second edition of MiC. Subcommittees for each content strand met to deliberate ways to strengthen the development of mathematical content within and across strands. Attending the meeting were several representatives from Holt, Rinehart and Winston (HRW), which just prior to the meeting had purchased the rights to market, sell, and distribute the materials. Since the HRW sales division is restricted to marketing textbooks to middle schools and high schools, a decision was made to focus the revision on a scope and sequence for grades 6–8. Given that the first edition of MiC involved forty units used across grades 5–8, the elimination of one grade level resulted in the reduction or elimination of topics that received limited attention in state and district curriculum frameworks (e.g., discrete mathematics). The attention that emerged over the previous decade on students' understanding of algebra also resulted in a decision to extend the algebra strand with a new grade 8 unit "Algebra Rules!" in addition to augmenting the entire algebra strand with a new supplemental resource, Algebra Tools. As with the first edition, the development team continued to hold editorial control of the mathematics content. However, when revising the blueprint for the second edition of MiC, the development team negotiated necessary compromises in the scope of mathematics addressed (the shift from a four-year to three-year program) and decisions regarding mathematical topics that required further development or reduced emphasis. These compromises were made for reasons that were both practical and market-driven.

Following the blueprint meeting, as with the first edition, the FI staff prepared revised or initial drafts of individual units. These drafts were reviewed by

American coauthors and underwent several iterations of revision between Dutch and American coauthors before the units were field-tested. For the second edition field test we were able to draw from a pool of experienced teachers who had used MiC for up to ten years in some instances. MiC teachers in several states field-tested units and, as part of the process, submitted students' work samples and periodic postinstruction reflections that focused on students' responses to new and revised content.

In contrast with the process used in development of the first edition, second-edition coauthors had limited opportunities to observe classrooms in person, and lessons were not videotaped. The development and field test of the first edition required significant investment in professional development to familiarize teachers with the RME approach to teaching, learning, and assessment; with an experienced group field-testing the second edition, further professional development was not needed. In place of extended classroom observations, unit coauthors depended on teachers' written reflections of students' responses to new and revised problem contexts, as well as on students' work samples. To enhance the quality of field-test data, surveys were embedded in teachers' guides, so that teachers could record productive and problematic lessons, examples of students' strategies, and suggested instructional modifications. Each unit also included "field test focus sheets" at crucial points in each unit, with prompts to support teachers' documentation of changes in students' access to mathematical content, teachers' identification of problems that were thought particularly revealing, and the emergence of students' representations and solution strategies. On the basis of their responses to field-test surveys and focus sheets, some teachers were invited to join the development team to produce lesson notes and instructional support material for the teachers' guides.

Lessons Learned and Future Implications

The development of Mathematics in Context through two editions involved a sustained, international collaboration between researchers at the Freudenthal Institute and the Wisconsin Center for Education Research. In addition to school districts in the United States, MiC has also been adapted for use in schools in Argentina, Canada, Denmark, Great Britain, India, Indonesia, Peru, and South Korea. Further examination of the implementations of MiC in different countries might offer a glimpse of the globalization of mathematics curriculum. Whether by positive results in students' achievement data, external evaluations of middle grades curricula, reports of improved classroom practice, or teachers' anecdotal reports of students' insight, collectively such evidence suggests that the design principles for Realistic Mathematics Education can be used successfully as guidelines for curriculum development.

References

Ball, Deborah Loewenberg. "With an Eye on the Mathematical Horizon: Dilemmas of Teaching Elementary School Mathematics." *Elementary School Journal* 93, no. 4 (1993): 373–97.

Bransford, John D., Ann L. Brown, and Rodney R. Cocking. *How People Learn: Brain, Mind, Experience, and School.* Washington, D.C.: National Academy Press, 1999.

Cobb, Paul. "Theories of Mathematical Learning and Constructivism: A Personal View." Symposium held at the "Trends and Perspectives in Mathematics Education" conference, Klagenfurt, Germany, February 1994.

Cobb, Paul, Ada Boufi, Kay McClain, and Joy Whitenack. "Reflective Discourse and Collective Reflection." *Journal for Research in Mathematics Education* 23 (May 1997): 258–77.

de Lange, Jan. *Mathematics, Insight, and Meaning.* Utrecht, Netherlands: Vakgroep Onderzoek Wiskundeonderwijs en Onderwijscomputercentrum, Rijksuniversiteit, 1987.

de Lange, Jan, and Freudenthal Institute Staff. *Framework for Classroom Assessment in Mathematics.* Unpublished manuscript. Madison, Wis.: National Center for Improving Student Learning and Achievement in Mathematics and Science, University of Wisconsin—Madison, 1999.

Dekker, Truus, Nanda Querelle, Margaret R. Meyer, David C. Webb, Mary C. Shafer, Jack Burrill, and Margaret A. Pligge. *Balanced Unit Assessments* (Grades 5–8). Chicago: Encyclopaedia Britannica, Inc. 2000.

Freudenthal, Hans. *Didactical Phenomenology of Mathematical Structures.* Dordrecht, Netherlands: D. Reidel, 1983.

Frykholm, Jeffrey. "Teachers' Tolerance for Discomfort: Implications for Curriculum Reform in Mathematics Education." *Journal of Curriculum and Supervision* 19, no. 2 (2004): 125–49.

Gravemeijer, Koeno. "Context Problems and Realistic Mathematics Instruction." In *Context, Free Productions, Tests, and Geometry in Realistic Mathematics Education,* edited by Koeno Gravemeijer, Marija van den Heuvel-Panhuizen, and Leen Streefland, pp. 10–32. Utrecht, Netherlands: Vakgroep Onderzoek Wiskundeonderwijs en Onderwijscomputercentrum, Rijksuniversiteit, 1990.

———. "An Instruction-Theoretic Reflection on the Use of Manipulatives." In *Realistic Mathematics Education in Primary School,* edited by Leen Streefland, pp. 57–76. Utrecht, Netherlands: CD-ß Press, 1991.

———. *Developing Realistic Mathematics Education.* Utrecht, Netherlands: Vakgroep Onderzoek Wiskundeonderwijs en Onderwijscomputercentrum, Rijksuniversiteit, 1994.

Her, Teresa, and David C. Webb. "Retracing a Path to Assessing for Understanding." In *Insight Stories: Assessing Middle School Mathematics,* edited by Thomas A. Romberg, pp. 200–220. New York: Teachers College Press, 2004.

Meyer, Margaret R. "Multiple Strategies = Multiple Challenges." *Mathematics Teaching in the Middle School* 4 (May 1999): 519–23.

———. "Representation in Realistic Mathematics Education." In *The Roles of Representation in School Mathematics,* 2001 Yearbook of the National Council of Teachers of Mathematics (NCTM), edited by Albert A. Cuoco and Frances R. Curcio, pp. 238–50. Reston, Va.: NCTM, 2001.

Meyer, Margaret R., Truus Dekker, and Nanda Querelle. "Context in Mathematics Curricula." *Mathematics Teaching in the Middle School* 6 (May 2001): 522–27.

National Council of Teachers of Mathematics (NCTM). *Curriculum and Evaluation Standards for School Mathematics.* Reston, Va.: NCTM, 1989.

———. *Assessment Standards for School Mathematics.* Reston, Va.: NCTM, 1995.

Romberg, Thomas A. "Problematic Features of the School Mathematics Curriculum." In *Handbook of Research on Curriculum,* edited by Philip W. Jackson, pp. 749–74. New York: Macmillan, 1992.

———, ed. *Standards-Based Assessment in Middle Grades Mathematics: Rethinking Classroom Practice.* New York: Teachers College Press, 2004.

Shafer, Mary C., and Thomas A. Romberg. "Assessment in Classrooms That Promote Understanding." In *Mathematics Classrooms That Promote Understanding,* edited by Elizabeth Fennema and Thomas A. Romberg, pp. 159–84. Mahwah, N.J.: Lawrence Erlbaum Associates, 1999.

van den Heuvel-Panhuizen, Marija. *Assessment and Realistic Mathematics Education.* Utrecht, Netherlands: Center for Science and Mathematics Education Press, Utrecht University, 1996.

———. *From Core Goals to Learning-Teaching Trajectories as a Guide for Teaching Primary-School Mathematics in The Netherlands.* Paper presented at the Twenty-sixth Conference of the International Group for the Psychology of Mathematics Education, Norwich, U.K., 2002.

van Reeuwijk, Martin, and Monica Wijers. "Students' Construction of Formulas in Context." *Mathematics Teaching in the Middle School* 2 (February 1997): 230–36.

Webb, David C. "Enriching Assessment Opportunities through Classroom Discourse." In *Standards-Based Mathematics Assessment in Middle School: Rethinking Classroom Practice,* edited by Thomas A. Romberg, pp. 169–87. New York: Teachers College Press, 2004.

Webb, David C., and Margaret R. Meyer. "Summary Report of Achievement Data for Mathematics in Context: A Connected Curriculum for Grades 5–8." Madison, Wis.: Wisconsin Center for Education Research, 2002.

Webb, David C., Thomas A. Romberg, Michael J. Ford, and Jack Burrill. "Teacher Collaboration: Focusing on Problems of Practice." In *Understanding Mathematics and Science Matters,* edited by Thomas A. Romberg, Thomas P. Carpenter, and Fae Dremock, pp. 231–51. Mahwah, N.J.: Lawrence Erlbaum Associates, 2005.

Webb, Norman, and Thomas A. Romberg. "Implications of the NCTM *Standards* for Mathematics Assessment." In *Mathematics Assessment and Evaluation,* edited by Thomas A. Romberg. Albany, N.Y.: State University of New York Press, 1992.

Wijers, Monica, Anton Roodhardt, Martin van Reeuwijk, Truus Dekker, Gail Burrill, Beth R. Cole, and Margaret A. Pligge. "Building Formulas." In *Mathematics in Context,* edited by Wisconsin Center for Education Research and Freudenthal Institute. Chicago: Encyclopædia Britannica, Inc., 2006.

7

The Case of

MathScape:
Seeing and Thinking Mathematically
in the Middle Grades

Glenn Kleiman
Emily Fagan
Susan Janssen
Amy Brodesky
Dan Tobin

THE design and development of the MathScape curriculum has taken many turns from when the original proposal was written in 1990 to the most recent revision, MathScape 2005, published by Glencoe/McGraw-Hill.[1] We learned that developing, publishing, and supporting a complete curriculum for grades 6–8 was far more complex than we could have imagined. Along the way, we encountered major changes in state standards and assessments, a series of publishers as each one was acquired by a larger company, and an increasingly adversarial national dialog about mathematics education.

Throughout the twists and turns over these fifteen years, our work has been guided by a vision of effective mathematics classrooms in which *all* students are successful in learning significant mathematics. Our vision encompassed many dimensions of teaching and learning mathematics: (1) students building their

The MathScape: Seeing and Thinking Mathematically curriculum was developed by the Education Development Center, Inc. (EDC), Newton, Massachusetts, and is published by Glencoe/McGraw-Hill. The Shell Centre for Mathematics Education (led by Hugh Burkhardt) at the University of Nottingham and EdMath Curriculum Services in Australia (led by Charles Lovett) made substantial contributions to the development of the curriculum.

1. The work on which this chapter is based was funded in part by the National Science Foundation (NSF) through Grant No. ESI-9054677 to Education Development Center, Inc. (EDC). Any opinions, findings, conclusions, or recommendations expressed here are those of the authors and do not necessarily reflect the views of the NSF.

knowledge of essential mathematics concepts, skills, and facts so they could apply this knowledge with fluency, flexibility, and deep understanding; (2) teachers taking on multifaceted roles by actively engaging students in rich mathematical investigations, facilitating discussions in which students communicate and clarify their mathematical ideas and assessing students' understandings; (3) meeting the developmental needs of middle school students by sequencing instruction to move from concrete to abstract and by helping students transfer from their own informal vocabulary to the more precise and powerful language of mathematics; and (4) engaging and motivating students through contextualized lessons, through creative and collaborative work, and through opportunities to share their own knowledge and experiences. (This vision is described further in Kleiman, Tobin, and Isaacson 1998.)

Our challenge was to design a curriculum that would bring this vision of mathematics into middle school classrooms throughout the country. At the same time, we set out to produce a curriculum that was highly teachable, one that reflects the day-to-day realities of middle school classrooms and provides the supports teachers need to be successful. The result was a curriculum with seven major units for each grade, along with a Math Handbook for review of basic concepts and skills and, in MathScape 2005, additional CD-ROMs and online support for teachers' lesson planning, classroom resources and assessments, and online activities for students.

Our project involved an interdisciplinary team, with staff and consultants who brought perspectives from research mathematics; teaching mathematics at the elementary, middle, and high school levels; ethnomathematics; teaching language arts and science; special education; developmental and cognitive psychology; curriculum development; educational software design; educational research; editing; and publishing. Most important, we also convened teams of teachers to brainstorm ideas, help develop the materials, review drafts, and field-test pilot versions of materials in their classrooms.

Design Principles

In developing the curriculum, we sought to bring together the project team's collective knowledge about mathematics content, middle school students, effective mathematics teaching, educational research, and curriculum design. Our goal was to develop creative, mathematically rich investigations that would reflect our vision of middle school mathematics classrooms and work well for a wide range of teachers and students. We sought to convey these activities in accessible and engaging materials for both students and teachers. In order to do so, we articulated a set of design principles that formed the foundation of our development work. These principles are summarized below.

Incorporate Three Components of Mathematics Content

We shared with the other NSF-funded projects the view that although the curriculum must contain the specific mathematics concepts and skills appropriate for middle school students, it should also contain more than the basic content typically seen in traditional materials. We also wanted to engage middle school students in what we considered the heart and soul of real mathematical work: *central mathematical ideas* that enable students to understand important connections across content areas and *mathematical habits of mind* that provide ways of thinking, understanding, and exploring with mathematical insight and precision.

Mathematical content strands. Along with the other NSF-funded projects, we used the NCTM *Standards* (NCTM 1989, 2000) as a guide to the specific concepts and skills to be included in the four major content strands of (1) number and operations, (2) algebra and functions, (3) probability and statistics, and (4) geometry and measurement. Within each strand, we developed curriculum units to address these concepts and skills from the NCTM *Standards,* with additions to address state and local curriculum frameworks. Our major challenge in this area was addressing the balance between the depth and the breadth of content. We sought to make sure that students had sufficient opportunities in the curriculum to master and apply important concepts and skills and that they had the opportunity to deepen their understanding of these concepts as they progressed through the sequence of units.

Central mathematical ideas. Our focus on four central mathematical ideas that connect specific concepts and strands helped us avoid the danger of producing a curriculum that would be "a mile wide and an inch deep." These central ideas are *proportional reasoning, multiple representations, patterns,* and *functional relationships.* For example, *proportional reasoning*—reasoning about relationships rather than quantities—is essential in understanding fractions, ratios, rates of speed, scale factors and rescaling, similarity, slope, geometric growth, and other important concepts in mathematics. In the curriculum, this central idea appears in many units and contexts. For example, students delve into the concept of proportionality when they solve problems with linear relationships in the "Patterns in Number and Space" unit, create scale drawings in the "My Travels with Gulliver" unit, explore probability in the "Chance Encounters" unit, and work with ratios in the "Roads and Ramps" unit. Similarly, the concepts of multiple representations, patterns, and mathematical functions are highlighted throughout the curriculum.

Mathematical habits of mind. The importance in our work of mathematical habits of mind was captured by the phrase *Seeing and Thinking Mathematically,* which was the title of our original proposal and is the subtitle of the published curriculum. We want students to actively engage in doing real mathematics, which involves processes such as abstracting, representing, pattern seeking, generalizing, conjecturing, experimenting, visualizing, seeking precision, using mathematical tools, systematizing, and developing proofs. Throughout the MathScape units, we furnish opportunities for students to develop and exercise these habits of mind. For example, in "Designing Spaces," students visualize and represent cube buildings from different perspectives; in "Patterns in Numbers and Shapes," students identify and extend visual and numeric patterns and make generalizations; in the "Language of Algebra" unit, students use multiple representations to compare different functions; in "Chance Encounters," students make and test conjectures about probability experiments; and in "Language of Numbers," they explore the properties of both real number systems and systems they create themselves. The important principle is that students should experience, at appropriate levels, the real work of mathematics and of applying mathematics in many fields.

Help Students Develop Meaningful Understandings of Mathematics

In developing MathScape, our goal was to enable students to go beyond rote procedures to develop meaningful understandings of important ideas. But what exactly are *meaningful understandings* and how do we know when students have arrived at them? Those were questions our project staff returned to repeatedly throughout the development of the curriculum. One analogy that proved useful was deciding between two ways for finding your way from one location to another: having a set of step-by-step instructions (e.g., turn left at the fourth traffic light, then right at the church) versus having a road map. The step-by-step instructions are analogous to the rote procedure: they may get you to your destination, but if you miss a turn or encounter a detour or a change in the situation, or if you want to reach your destination from another starting point, they are useless. The map, analogous to a meaningful understanding, enables you to deal with changes in the situation, identify and correct errors, find alternative routes to your destination, and plot routes for entirely different starting and ending points.

Within our curriculum units, we include problem sets that require students to go beyond a simple rote procedure to use multiple approaches to solving complex problems, to generalize approaches to related problems, to articulate and share their approaches, and to find connections between new ideas and their prior knowledge. Across units, concepts get generalized and deepened through their use in a variety of forms and contexts. We viewed rote procedures applied without understanding as a type of mathematical misconception, and we looked for ways to help teachers recognize when students were not able to use mathematics with understanding. We also helped teachers and students understand and correct other common misconceptions about important mathematical ideas (e.g., "multiplying always makes numbers larger"). Throughout the curriculum, we include From the Classroom tips to teachers in order to suggest ways to address students' misconceptions. We created a powerful activity called "Dear Dr. Math," in which students answer a letter from someone who is puzzled about a mathematical idea and is wrestling with a common misconception. In addition, MathScape assessments tap students' abilities to flexibly use what they have learned and to communicate the rationale for their approaches in order to help teachers see if students have obtained meaningful understandings.

Place Mathematics Learning within Contexts Relevant to Middle School Students

This design principle was based on our understanding of how we could best engage middle school students in "seeing and thinking mathematically." We knew, from both researchers and practitioners, that many middle school students lose interest in mathematics. When we asked, "What interests and engages students that can be incorporated into the curriculum?" we found that students want to "see a purpose in what they are learning" and to "figure out how the world works and their place in it." We combined this perspective on middle school students with lessons from ethnomathematics to articulate a view that framed the start of our original proposal to NSF:

> To be human is to seek to understand. Mathematics, along with science, has made possible dramatic advances in our understanding of the physical universe. To be human is to explore. Throughout history, mathematics has been essential for explorations, from navigating by the stars to traveling into space. To be human is to participate in a society. Societies require mathematics in order to keep records, allocate resources, and make decisions. To be human is to build, and mathematics is essential for the design and construction of everything from tents to temples to skyscrapers. To be human is to look to the future. Mathematics enables us to analyze what has been, predict what might be, and evaluate our options. To be human is to play, and mathematics is part of our games and our sports. To be human is to think, to create, and to communicate. Mathematics provides a vehicle for thinking, a medium for creating, and a language for communicating. Indeed, to be human is to develop mathematics. Mathematics has been developed in every culture for the purposes of counting, locating, measuring, designing, playing, and explaining.

This perspective led us to build the curriculum as a series of context-based units, with the contexts providing a framework for significant mathematical investigations. Over time, we evolved a very broad range of contexts. Some were based on real-world activities, such as designing homes in "Designing Spaces: Visualizing, Planning, and Building"; some were based on primarily mathematical contexts, such as "The Language of Numbers: Inventing and Comparing Number Systems"; some were based on problem-solving puzzles, such as "Patterns in Numbers and Shapes: Using Algebraic Thinking"; and some on game contexts, such as "Chance Encounters: Probability in Games and Simulations." Others were based on literary contexts, such as "Gulliver's Worlds: Measuring and Scaling." Within each context, we looked for opportunities to connect the context with important mathematical ideas, concepts, and processes. A few examples include using functions to predict population growth, learning about probability to design fair games, creating representations of three-dimensional forms to create building plans, using proportions to make cost comparisons, and learning about measurement and scale to make two- and three-dimensional objects at the actual sizes they would be in the worlds Gulliver visited.

To cover all the essential topics, we found it necessary to include some units focused on specific mathematical topics, such as fractions, which are less context-driven than others. But in each unit, our goal was to convey to students the view of mathematics as central to the human experience as opposed to what we considered a "dehumanized and decontextualized" view of mathematics that is too often predominant in grades K–12 mathematics classes.

Offer Varied Learning Opportunities to Support Diverse Learners

In order to design a curriculum that could enable all middle school students to succeed in mathematics, we needed to address diversity in many senses of the word, including diversity in learning styles, mathematical backgrounds, cognitive strengths and weaknesses, and cultural and linguistic backgrounds. We also had to address diversity among the teachers in their knowledge of mathematics and preferred approaches in the classroom. We therefore included a varied set of activities in the curriculum, seeking to provide alternative opportunities for students to succeed and a rich set of resources that teachers can adapt for their own students. Some of our specific principles in working toward this goal included the following:

- Design activities so that all students, regardless of their abilities, would be able to get started and have some success. We aimed to develop activities that could be approached in multiple ways and at different levels of sophistication, so that teachers could find appropriate variations to fit their students.

- Build students' mathematical understandings from the informal to the formal, from the concrete to the abstract, and from the students' own language to systematic mathematical representations. Different students will be ready to progress to different levels of these steps from informal, concrete representations of the pattern to formal, generalizable mathematical descriptions, but all middle school students should be able to have some success.

- Include activities that would enable different students to make use of their individual learning strengths. Throughout the curriculum, we provided investigations involving physical models, kinesthetic experiences, visualization, writing, discussion, collaboration, and symbolic work, in order to meet diverse learning styles and multiple intelligences.

- Furnish opportunities for students to work collaboratively in activities that called for different strengths and roles, so that each student could contribute and students could learn from one another.

- Provide, whenever possible, activities that could benefit from cultural diversity in the classroom by designing *culturally open activities.* In these activities, we asked students to contribute their own examples to enrich the range of possibilities considered in the lessons. For example, in "Chance Encounters," students bring in examples of games from different societies, both current and historical, to analyze their statistical properties.

Support the Teacher

The MathScape curriculum was designed with the view that the success of mathematics education depends on the interaction of the teacher, the students, and the curriculum materials. Only the teacher knows the individual students, can monitor their day-to-day progress, create a productive classroom atmosphere, engage students' interest, and adjust lessons to his or her own students. We viewed ourselves as the teacher's consultants, offering ideas and recommendations for the teacher to evaluate and use as appropriate for his or her students. We aimed to provide extensive support for teachers, knowing that, in many instances, the curriculum was asking them to teach new content and to employ new approaches in their classrooms.

In order to help teachers use the curriculum to meet their students' needs, we included two features that were generated by field-test teachers: Teacher Reflections essays about teachers' experiences with a unit, and From the Classroom tips that offered suggestions for implementing a lesson in the classroom, often including ways to help struggling students. Both were designed to provide images of the lessons in action in real classrooms and to convey to teachers that they should implement activities with some flexibility, making choices based on the needs of their students. We also included examples of students' work to help convey how the curriculum will work in the classroom.

In addition, the lesson plans and teacher support materials were designed to furnish the following supports for teachers:

- To help teachers understand the central mathematical ideas of the unit, the curriculum outlines mathematical goals and offers background information and explanations.

- To help teachers engage students in mathematically rich investigations and discussions, each lesson contains a suggested set of steps and detailed guidance for less-experienced teachers.

- To guide teachers in facilitating classroom discussions, each lesson contains sample discussion questions, examples of typical responses by students, and other suggestions for leading effective discussions.

- To help teachers assess students' learning, the curriculum provides a variety of tools, including embedded assessments, skills quizzes, final projects, and accompanying rubrics for assessing students' performance.

- To help teachers communicate with parents, administrators, and other teachers, each unit contains overviews and sample letters to parents.

- To help teachers meet the needs of diverse learners, the curriculum contains suggestions about different ways to adapt lessons for different student needs, including supplying challenging extension activities for advanced students.

- To help students review prerequisites, practice skills, and prepare for tests, materials to support these needs are provided in a Math Handbook.

- To help teachers plan lessons, provide practice, make adaptations for students with special needs, and customize assessments, a TeacherWorks CD tool was created for the 2005 revised edition.

Design for Robustness in Real Classroom Settings

It's easy to imagine developing a curriculum for an idealized context—teachers fluent in all the mathematics and experienced with all the pedagogical approaches, students who have already mastered all the prerequisites, sufficient class time devoted to mathematics, easy access to materials and technologies, support at home for study and homework, and so on. However, a curriculum must be designed to offer useful guidance and resources across a wide variety of contexts, many of which are far from meeting these idealized conditions. Therefore, the challenge is to design mathematics lessons that are robust and that will be usable and effective even in settings that are far from ideal—with teachers who have limited knowledge of mathematics, with students who had poor mathematics instruction in prior years, and in classrooms that are short of materials, technologies, and time.

Development Process and Lessons Learned

Far more was involved in the development process than can be summarized here. Like the curriculum itself, the process continually evolved according to the task at hand, the project timeline, the makeup of the development team for a given unit, and the lessons we learned along the way. Here are a few of the things we wished we had known at the outset of the project.

Many Steps, Many Contributors

The outline of the development process is deceptively simple: draft a unit, field-test it, and then revise it on the basis of the test results. In actual practice, there are many more steps, each of which requires successful collaboration among people who bring different perspectives and expertise. The MathScape development process, in abbreviated form, included the following:

1. As background to the initial planning of the curriculum, staff members reviewed mathematics content objectives in the NCTM *Curriculum and Evaluation Standards,* state frameworks, and other middle school curricula. We also reviewed relevant educational research on middle grades mathematics content, particularly students' misconceptions and strategies for addressing them. Finally, we gathered feedback from middle school teachers about areas of difficulty for their students and about effective practices.

2. Staff and consulting teachers engaged in brainstorming sessions to generate ideas that brought together significant mathematics content with contexts and activities that would engage middle school students. These were stimulating sessions that resulted in an abundance of ideas that then had to be filtered and sorted to select those that would be incorporated into specific units.

3. Important content was outlined and contexts identified to go into individual units, with the recognition that units often took on a life of their own during the development process and that the content and contexts were likely to evolve.

4. A development team of two to three people created an initial set of ideas for *foundation* activities that would provide the main scaffolding for a particular unit. The idea was to design and test some primary investigations centered on mathematical goals and, once those were in place, build the rest of the unit around them. The development team would take initial ideas into classrooms to try out with groups of students to ascertain whether the important mathematics was coming through and whether the context was engaging to students.

5. On the basis of the results of these initial tests, some ideas were dropped, some refined, and others moved forward with a great deal of energy, since they seemed to really work in the classroom. The development team then wrote the first draft of a unit. This draft would include all activities, background information for the teacher, and assessment tools. We also added a very helpful role of "core reviewers" for units. These were staff or consultants who were not participating in writing the unit but who were asked to be readily available to serve as consultants and to prepare very detailed reviews of the draft lessons.

6. Next came a classroom field test with a set of teachers who would agree to teach the entire unit in place of a part of their standard curriculum. Prior to beginning work in the classroom, the developers met with a team of field-test researchers to define important questions about the unit. The researchers then observed two to three times a week in local classrooms for the five to six weeks of the test, collected students' work, and interviewed teachers on a weekly basis. The researchers also conducted phone interviews with field-test teachers in schools in other parts of the country.

7. At the same time that the units were tested in classes, they were also sent to outside reviewers, including consultants, university mathematicians, and other EDC mathematicians and mathematics educators. These reviewers checked the materials for mathematical accuracy and offered suggestions for improving the lessons.

8. After the completion of the first field test and the outside reviews, the developers and field-test researchers would debrief for one to two days about the results of the field test, going through each lesson in detail, discussing what worked and how teachers and students responded. They would then work through specific revisions, which might range from clarifying directions for students to changing the sequence of lessons to major rewriting. In addition, developers and field-test researchers would identify important points to include in the From the Classroom tips to teachers and would help selected field-test teachers complete their narratives for the Teacher Reflections pieces.

9. The developers would create a revised version of the unit in response to the recommendations from the field test and from the outside reviewers. In some instances, the revisions were so substantial that the entire field-test process was repeated with the revised version.

10. The development team worked on strengthening the sequence of units within each grade and across the grades. They strove to address all the mathematical content needed for each grade; to make sure important content appeared an appropriate number of times across units; to balance the number of large projects, group activities, hands-on activities, writing activities, and so on, across the units for the grade; to supply all the necessary homework, supplemental, and extension activities; and to make sure everything was as clear and succinct as possible for both students and teachers.

As we began to work with the publisher, another round of revisions took place to fit publication constraints (e.g., the total number of pages, the number of re-producibles, the types of hands-on materials, etc.) and to incorporate the views of the publisher about what makes a curriculum that is not only usable but also marketable.

Developing a Shared Vision

As we noted above, we had an interdisciplinary development team, with many perspectives represented. We believe this strengthened our work, but it also raised many challenges, particularly in the early years of the project. We grappled with creating a shared vision on what was most important, what defined a high-quality unit, how to best guide teachers and students, and all the rest. Building a shared vision to enable the work to move forward productively took a great deal of time, debate, working collaboratively on lessons and units, observing classrooms together, listening, coming to common understandings of important terms, and learning together. If we could start again, we would plan far more time and specific project activities to foster developing a shared vision early in the work, and then repeat this process when we began working with the publisher.

Parts and Wholes

Which comes first—a given unit or a map of the entire curriculum? At times, it seemed as though the development team could not move forward on specific units without first defining the structure of the whole curriculum, but it was equally difficult to define the whole structure without agreement on the format and style of some of the individual parts. We never fully resolved this top-down versus bottom-up tension, and ended up with an approach that moved back and forth between the parts and the whole. Working from the NCTM *Curriculum and Evaluation Standards* as a content guide, we built some initial units, using those to establish our models for lessons and units. We then began to place them into the overall framework, defining what needed to be in the future units. As more units were developed, we further articulated the content across the curriculum, and we made adjustments to individual units to balance overall content coverage and ensure a strong flow across the units. It would have been simpler to work from tight overall guidelines at the start, but it also would have limited the creativity of the materials, which was very important to the development team.

Field Testing

We consistently found that no matter how many people, including current teachers, were involved in developing the materials, there were always surprises when we took them out into real classrooms, and that extensive field testing is essential to successful curriculum development. Our field testing included trial lessons during the development process, in-depth testing of draft units in a small number of classrooms, and distance field tests with less-detailed feedback from teachers. All levels of field testing yielded important results, but we found the most valuable information came from the in-depth testing during which our formative research team observed classrooms, interviewed teachers and students, and analyzed students' work. The debriefing sessions in which the formative research team worked with the curriculum developers, going through a unit lesson by lesson and step by step to see what worked well and what needed to be improved, were invaluable.

Seeking the Right Balance

There are many demands on a curriculum. Throughout the process, we struggled with a complex balancing act, weighing such issues as depth versus breadth of content; structure versus flexibility of activities; individual, small-group, and whole-class activities; context-driven and content-driven activities; coverage across the content areas; focus on specific content versus central ideas versus mathematical processes; detailed explanations versus the need to be succinct; time allocated to different parts of the lesson (e.g., introducing activities, engaging students in activities, and follow-up discussions and reflections); types of activities (visual, kinesthetic, computational, etc.); suggestions for advanced students and for struggling students; and publishing and page-limit constraints. Although our design principles and field-test results provided some guidance, we found that there were no simple answers: finding the right balance is the art of successful curriculum development.

One Piece of the Puzzle

From the beginning of the MathScape project, we viewed curriculum development as just one piece of the puzzle of mathematics education reform. We knew that major efforts were under way to improve the preparation and professional development of teachers, develop new types of student assessments, and furnish information to administrators, parents, and others about changes in mathematics education. We viewed making our materials synergistic with these efforts as essential to our contributing to the improvement of middle school mathematics teaching and learning. We have continued our work with districts on effective implementation and professional development strategies through the Show Me Middle School Curriculum Implementation Center, which was funded by NSF through 2005. We believe this type of support is essential, since curriculum reform can be successful only when it is one piece of an integrated, well-planned mathematics education improvement effort. That is an essential lesson learned, no matter what type of curriculum is to be implemented.

A Final Note

We were privileged to have the opportunity to work in collaboration with so many dedicated educators to develop an innovative product that is both engaging and mathematically rigorous and to be able to support its use in schools for many years following its publication. We have been rewarded by its successful use by many teachers and students and by the opportunity to continue our work with the new MathScape 2005 edition. We are pleased to have been able to contribute to the national discussion about the important task of improving middle school mathematics education. At the same time, we are dismayed by the often adversarial nature of the discussion about mathematics education in the United States and by the fact that many teachers and students have been affected by shifting demands and pressures that have often seemed to be driven more by political than by educational concerns. We hope we can contribute to a more productive dialog, one that recognizes, as we learned while developing the MathScape curriculum, that there are no simple answers, that the challenge is finding a good balance of the many desirable elements in a mathematics curriculum, that a curriculum is just one piece of the complex puzzle of improving mathematics education, and that a great deal is yet to be learned about teaching children mathematics.

REFERENCES

Kleiman, Glenn M., Dan Tobin, and Shelley Isaacson. "What Should a Middle School Mathematics Classroom Look Like? Watching the 'Seeing and Thinking Mathematically' Curriculum in Action." In *Mathematics in the Middle,* edited by Larry Leutzinger, pp. 120–28. Reston, Va.: National Council of Teachers of Mathematics, 1998.

National Council of Teachers of Mathematics (NCTM). *Curriculum and Evaluation Standards for School Mathematics.* Reston, Va.: NCTM, 1989.

———. *Principles and Standards for School Mathematics.* Reston, Va.: NCTM, 2000.

8

The Case of

MATHThematics

Rick Billstein
Jim Williamson

I N 1992, the National Science Foundation (NSF) awarded a grant to the University of Montana to design, develop, and evaluate an innovative mathematics program for the middle grades that would reflect the recommendations of the National Council of Teachers of Mathematics (NCTM) *Standards* documents (1989, 1991). The Six Through Eight Mathematics (STEM) project achieved those goals, and the resulting curriculum was published by McDougal Littell, Inc., under the title *Middle Grades MATHThematics* (Billstein and Williamson 1999).

Design Principles

The goal of the STEM project was to develop a curriculum that would help all students develop their abilities to

- reason logically;
- apply mathematics to real-life activities;
- communicate about and through mathematics;
- make connections among mathematical concepts and relate them to other content areas;
- use quantitative and spatial information and problem solving to make decisions;
- become independent learners who are well prepared for the real world and for future mathematics courses.

Mathematical Content

The mathematics content of the curriculum was selected with the aim of involving students in doing mathematics that was new, interactive, meaningful, and interesting. The content was organized around two components: content strands and unifying concepts.

Content strands. To ensure that a complete three-year curriculum (grades 6–8) was provided and that it was broad enough and balanced enough at each grade level, the learner outcomes were organized in strands: Number, Measurement, Geometry, Statistics, Probability, Algebra, and Discrete Mathematics. However, the goal was that the development of the content within each strand would (*a*) emphasize problem solving, critical thinking, and reasoning over rote procedural drill; (*b*) decrease the attention given to the review of "elementary" topics such as whole-number computation; and (*c*) increase the emphasis on data analysis, proportional reasoning, algebra, geometry, and discrete mathematics.

Unifying concepts. Four unifying concepts—Proportional Reasoning, Multiple Representations, Patterns and Generalizations, and Modeling—were identified and used throughout the curriculum to help students develop mathematical concepts and make connections among mathematical ideas and between mathematics and the real world.

Proportional reasoning is the ability to express one number as a certain multiple of another. It forms the basis for understanding the concepts of ratio, rate, percent, proportions, slope, similarity, scale, linear functions, and probability.

Multiple representations are used to connect topics such as (*a*) coordinate systems and functions, (*b*) fraction-decimal-percent representations, and (*c*) geometric representations of arithmetic concepts. Exploring different representations of concepts helps students understand mathematical ideas by letting them make connections among concepts and by providing for different learning styles.

Identifying and describing numeric and geometric *patterns* and making, testing, and applying *generalizations* about the data gathered from problem situations are the tools students use to develop algorithms and construct mathematical meaning.

Modeling is the tool students use to connect mathematics to the real world. It is the process of taking a real-life problem, expressing it mathematically, finding a mathematical solution, and then interpreting the solution in the real-world context.

Organization and Instructional Approach

The STEM vision was that middle school should be a time when students are actively involved in doing mathematics that is new and meaningful, not just a period of review. To achieve this vision, the instructional approach was designed to involve students in doing mathematics. Throughout the STEM materials, students are actively engaged in

- investigating, discovering, and applying mathematics;
- using concrete materials to explore mathematical properties and relationships;
- working cooperatively;
- communicating ideas orally and in writing;
- using technological tools when appropriate;
- integrating mathematical strands to solve real-life problems.

We decided that the most effective way to achieve the goal of the STEM project was to present the mathematics content in thematic modules that develop mathematical ideas in relevant and meaningful contexts. The mathematics content of each module would be integrated and include cross-curricular connections to other content areas such as science, language arts, and social studies.

Each module would consist of four to six *Sections*, each requiring one to three days to complete. Each section would be further divided into *Explorations* where students would be actively involved in learning new mathematics. Explorations would be completed by students working individually, in cooperative groups, or as an entire class. The activities in the explorations would range from guided discovery to open-ended investigations.

Technology

When development of the STEM curriculum began, students' access to graphing calculators and computers was limited. However, we furnished graphing calculator, spreadsheet, and other software activities that teachers and students could use. Many of these activities were included in the texts; others were included in the Teacher's Resource Book for teachers to use as they saw fit. The idea was to make the program calculator-dependent and computer-enhanced. We also included a Web site with Web-based links to offer the latest information on the themes used in the modules. We supplied interactive Web-based activities that were aligned with each module. A CD for students with tutorial software and a CD for teachers with a test and practice-sheet generator were provided with the published program.

Assessment

In the STEM curriculum, assessment was envisioned as an integral part of the instructional process rather than as an add-on to it. Not only would assessment information be drawn from instructional tasks, but also the assessment tools themselves would help students master concepts and develop skills. The primary goal of the STEM assessment package was to improve learning. The assessment package would serve four major purposes:

1. To monitor students' progress in problem solving, reasoning, and communication
2. To assess students' proficiency in content areas
3. To help teachers make instructional decisions
4. To document students' progress for students, parents, and teachers

Assessment tools. Before we piloted the STEM materials, it became apparent that teachers needed guidance on how and when to interact with students. We needed ongoing assessment tools to supply information that teachers could use to decide what instruction is necessary to help students achieve the outcomes of the curriculum. The results of ongoing assessment would identify a variety of student needs, such as reteaching certain concepts, reviewing or practicing skills, or presenting additional material using a different model or teaching technique. To accomplish this aim, Checkpoints, Discussion Questions, and Try This as a Class exercises were included in the materials. Checkpoints are questions or problems that are used by the teacher to check students' understanding of a concept or skill before the students continue with the exploration. Discussion Questions provide places where students can check their understanding of a concept by sharing or generating ideas within their group or class. Try This as a Class exercises appear at points where direct instruction is needed to summarize important ideas or to bring closure to a line of inquiry. They are similar to Discussion Questions, except that the teacher directs the discussion or activity and guides the learning.

The STEM materials include Extended Explorations, or E^2s, that are typically open-ended or open-response problems that students complete independently—or occasionally in small groups—outside of class. E^2s involve a variety of mathematical concepts and may be solved in different ways. Each module contains one E^2. It usually takes one to two weeks for students to complete an E^2, and their solutions are assessed using an assessment scale, which is a multidimensional, generalized assessment rubric.

The STEM project believed that the key to raising students' performance is to actively involve students in assessing their own work. This was achieved through the use of Student Self-Assessment Scales. The scales were designed to help students answer the question, "How can I improve my problem-solving, reasoning, and communication skills?" Five scales were developed: Problem Solving, Mathematical Language, Representations, Connections, and Presentation. Combined, they provided a generalized rubric that defines the dimensions of mathematical investigation. As students become familiar with the scales, they understand what they need to do to improve their problem-solving, reasoning, and communication skills.

Teachers assess students' work using the same scales written from a teacher's point of view. The assessment scales allow the teacher to recognize a student's strengths and help the student focus on areas where improvement or growth is needed. The combination of student and teacher assessment offers important feedback to help students improve.

Portfolios were used to provide comprehensive documentation of students' progress in, attitude toward, and understanding of mathematics over a period of time.

Development Process

Many of the STEM writers were classroom teachers who could provide a reality check and discuss how the materials would work with students. The writers were chosen in a national search and came from all over the United States. These teachers were joined with outstanding mathematics educators from mathematics departments and schools of education and mathematicians from departments of mathematics.

The STEM curriculum modules were developed using an iterative process in which each module underwent at least three stages of field testing and revision. During their development, the curriculum materials and instructional strat-egies underwent five years of extensive field testing by more than 250 teachers in twenty-five states with more than 35,000 students. An outside evaluator chose sites to assess the effectiveness of the curriculum materials in different types of schools—urban, suburban, and rural—as well as in a variety of classroom set-tings. Project staff made frequent visits to the field-test sites to observe classes and to consult with teachers. Formative evaluation data were collected from the sites by the evaluator, and the teachers met in the summer to give feedback on materials and to prepare for the use of upcoming modules. The information sup-plied by the evaluator and by the teachers involved in the field tests was used to revise, edit, and rewrite the materials. This close collaboration of the curriculum writers with teachers at diverse school test sites was a crucial factor in produc-ing materials that work in typical American mathematics classrooms—for both students and teachers.

Summative evaluation data were also collected. Results of this research and other outside research are detailed in Billstein and Williamson (2003). Further discussion of research on the STEM materials is included in two booklets: *Mc-Dougal Littell Middle Grades MATHThematics: Research-Based Framework* (2005) and *McDougal Littell Middle Grades MATHThematics: Learner Verifica-tion Studies* (2005).

During the development of the STEM curriculum, we encountered several bar-riers that affected our design principles. Some of the barriers were a result of pres-sure from our publisher and other outside sources, and others were unanticipated consequences of our project design.

Because the final curriculum had to be marketable nationwide, we had to try to meet the curriculum standards and frameworks of every state. Since there is little commonality in the mathematics content or the sequencing of topics at each grade level within these documents, this requirement made it very difficult to reduce the amount of the repetition of topics in the curriculum and to provide greater depth of coverage at each grade level. In short, the lack of a generally accepted consis-tent set of standards is a major factor in the watering down of the mathematics curriculum nationwide.

While developing the instructional materials, the STEM writing team respond-ed to ideas and experiences from many classroom and field-test teachers. As a re-sult, the teacher support materials for the MATHThematics modules are far more extensive than those of more conventional textbooks, and teachers report that they are very useful. However, producing the student materials was more costly than the publisher had anticipated, and as a result, significant compromises had to be made in the design of the teacher resources. The net result was that although the teacher resources are very good, they are not in a teacher-friendly format, and this significantly reduces their effectiveness.

As originally envisioned, the STEM curriculum is composed mainly of inqui-ry-based thematic activities. By its very nature, this approach requires a great deal more reading than what is typically found in a mathematics textbook. Our initial assumptions about the students' and the teachers' abilities to deal with this added dimension proved to be unrealistic. Not only did we have to revise the material several times to reduce the amount of reading and the reading level, but we also had to produce support materials to offer strategies teachers could use with students to help them with content area reading.

A major barrier in the first several field tests was that we did not receive adequate feedback on many of the modules. This was a direct result of the new pedagogy and the amount of reading in the materials. We assumed that teachers and students could complete the modules much faster than they did. In reality, most teachers were able to complete only half the modules, and the teachers who did more than half did not cover them completely. Because this new *Standards*-based thematic approach took more time than anticipated, we had to rewrite the materials several times to streamline the presentation.

Another problem that contributed to the poor completion rate was that we made unrealistic assumptions concerning the amount of practice students would require. We believed that because of the constructivist approach, students would require fewer practice activities. The results from the pilots and field tests revealed that this assumption was not entirely correct. We had to work closely with the teachers to find an appropriate solution to this problem.

The application-based teaching strategy used in MATHThematics encourages students to become actively engaged in the discovery of mathematical ideas. However, in many instances, field-test teachers did not have the mathematical background to develop the activity appropriately. Many of the field-test middle school teachers were elementary certified and had little mathematics training. These teachers required comprehensive professional development programs and enhanced teacher materials to articulate the important ideas being developed and the connections among the different explorations. We found that teachers needed to see how they could embed skill practice in a problem-solving activity without having to pass out supplementary drill-and-practice sheets.

What We Have Learned

The careful development and extensive field testing of the STEM materials along with extensive dissemination and implementation efforts led to an impressive number of MATHThematics adoptions. The information gathered from the field testing and the adoptions revealed a variety of ways in which the program can be improved. These include, but are not limited to, changes in mathematics content and approaches to concept development, the revision of the teacher-support materials, and the inclusion of materials to inform parents of the changes occurring in mathematics education and furnish them with tools to help their children.

Several NSF-funded elementary and high school curricula have now been published. As a result, students are beginning to enter middle school with different backgrounds, and our materials must reflect and take advantage of this change. At the other end, the new *Standards*-based high school programs are looking for better-prepared students coming into their programs. This is all part of an efficient and seamless K–12 curriculum consistent with the NCTM *Standards* documents (1989, 1991) as envisioned when this curriculum-development process was instituted in the early 1990s.

In addition, we have received a tremendous amount of teacher feedback on the commercial materials. It is important to evaluate this information and to make appropriate revisions based on it. We believe that several areas can be improved.

Rational numbers. Middle-grade teachers and parents are concerned about the coverage of rational-number topics, especially computational procedures involving fractions. The MATHThematics materials carefully develop the concepts of fractions, decimals, percents, and proportionality and relate them to real-world contexts. However, in light of the publication of NCTM's *Principles and Standards for School Mathematics* (2000), it is important to revisit this important topic and, if necessary, revise our treatment.

Algebra. The MATHThematics curriculum gives extensive attention to algebra and functions. Many schools that have adopted the MATHThematics curriculum have accelerated programs that move some of their eighth graders directly into high school courses in geometry (bypassing ninth-grade algebra) or into integrated mathematics programs such as those designed by the NSF-funded high school projects. Other schools find that their students are extremely well prepared for ninth-grade algebra or the first course in an integrated program. National pressures to raise mathematical expectations for all students have often focused on accelerating the high school curriculum by making algebra an eighth-grade course. We must examine the new NSF high school programs to see if there are content expectations that are not being covered, and if so, we must decide how to cover them. Schools that are using MATHThematics or are considering using it are concerned about whether an integrated approach is right for them in the eighth grade.

Technology. Data analysis and geometry utility software and graphing calculators with new capabilities are becoming more widely available. Several new products such as Tinker Plots will soon be added. It is essential that the use of technology on the part of MATHThematics be continually reexamined in light of new technology and new research and that the text materials be revised on the basis of the findings.

Learning in context. Research shows that mathematical meaning plays a vital role in students' solutions of problems in everyday activities, especially compared to in-school problem-solving activities that depend more on algorithmic rules. The MATHThematics curriculum makes extensive use of problem-based teaching approaches, often presenting students with mathematical ideas embedded in real-life application situations. Although this style of curriculum and teaching is exceptionally engaging for students, it also demands frequent updating to make sure the problem situations, data, and questions are indeed authentic in the critical eyes of young adolescents. The curriculum also has numerous ties to literature and science, and these need to be frequently evaluated to determine how well they are working and to see if there are more appropriate readings or activities.

Professional development. We have found that an enormous amount of professional development is necessary for successful, sustained implementation of this type of program and that a substantial revision of the professional development components of the program is required to adequately address this need. Research shows that one-time in-service workshops are unlikely to produce either significant or long-term change in the beliefs of mathematics teachers, their attitudes, their teaching methods, or their mathematical understanding. What is needed is a well-planned professional development component with follow-up discussions and meetings.

Parent materials. We recognize that parents play a major role when it comes to the education of their children. We need to develop a comprehensive plan to inform parents of the changes occurring in mathematics education and the rationale for the changes. We also need to provide parents with tools to help their children. This includes designing a set of activities that parents can use at home with their children, preparing a list of Web sites that can be used at home, and developing a video for use with parents.

Accessibility materials. The Individuals with Disabilities Education Act (1997) helps to ensure that students with disabilities receive the best possible mathematics education. We need to offer resources that will help schools ensure that middle school students with physical, sensory, cognitive, and psychosocial disabilities are able to participate and achieve in a MATHThematics classroom.

Concluding Comments

In summary, the design of the MATHThematics curriculum distinguishes it from other more conventional texts and from other NSF-funded textbooks. The thematic organization of modules is designed to stimulate the study of mathematics by grounding it in applied contexts. Each thematic module is intended to last four to six weeks and then a new theme is begun. The switching of themes keeps students interested in new contexts. The thematic approach has the potential for interdisciplinary study, making sensible connections among mathematical ideas and between mathematics content and other subject areas. When students see the mathematics taught in a thematic module, they become more interested in the mathematics and remember it longer. In the STEM curriculum, the mathematics is taught and practiced in a series of "explorations." Explorations are structured and lead students to a specific concept or procedure; instruction uses both student investigation and direct instruction to promote students' learning. The curriculum uses an integrated approach instead of covering only one content strand in a unit or chapter. In this way students see mathematics as a whole rather than as disjoint strands. Through its breadth and richness, the thematic approach lends itself to mathematical treatments from across the major mathematical strands. The integrated approach offers the opportunity to make connections among mathematical ideas throughout the curriculum as a natural outgrowth of students' work. The design and role of assessment was always an important feature in the development of the project. The multidimensional generalized scoring rubric for teachers along with the self-assessment rubric for students is unique to the STEM curriculum and is built in as an integral design feature and not as an add-on to the program. The rubrics are designed to help students answer the question "How can I improve my problem-solving, reasoning, and communication skills?" These design features have been well received by the mathematics education community nationally and internationally, and we will continue to revise the materials on the basis of these successful design features.

REFERENCES

Billstein, Rick, and Jim Williamson. *Middle Grades MATHThematics* (Books 1–3). Evanston, Ill.: McDougal Littell, 1999.

———. "Middle Grades MATH*Thematics:* The STEM Project." In *Standards-Based School Mathematics Curricula: What Are They? What Do Students Learn?* edited by Sharon L. Senk and Denisse R. Thompson, pp. 251–84. Mahwah, N.J.: Lawrence Erlbaum Associates, 2003.

McDougal Littell Middle Grades MATHThematics: Learner Verification Studies. Evanston, Ill.: McDougal Littell, 2005.

McDougal Littell Middle Grades MATHThematics: Research-Based Framework. Evanston, Ill.: McDougal Littell, 2005.

National Council of Teachers of Mathematics (NCTM). *Curriculum and Evaluation Standards for School Mathematics.* Reston, Va.: NCTM, 1989.

———. *Professional Standards for Teaching Mathematics.* Reston, Va.: NCTM, 1991.

———. *Principles and Standards for School Mathematics.* Reston, Va.: NCTM, 2000.

Part 3

Design and Development of Grades 9–12 *Standards*-Based Curricula

9

The Case of the

CME Project

Al Cuoco

C URRICULUM design in U.S. precollege mathematics is largely topic-driven; a course is defined by the topics it treats. Major criteria for selecting topics in any particular course include the following questions:

- Does this topic review and deepen important ideas from previous courses?
- Is it a prerequisite for likely subsequent courses?
- Did it fall through the cracks in earlier grades?
- Is it on the state test?

As one moves up the grades, the effects of this design principle (selecting the topics) compound. By the time one reaches the high school level, the result is a huge compendium of topics that range from graphing equations to triangle trigonometry to data analysis to complex numbers. These monster texts have become de facto definitions for the American high school mathematics curriculum.

Of course, there's much more in each of these texts than what students will ever need for future courses and what teachers can possibly teach in a year. Indeed, it's a well-known fact among high school teachers that one can finish only slightly more than half of the chapters in these texts in a given year, and yet many students who go on to the next course from such experiences do have what they need to get respectable grades.

The quality of a curriculum depends, more than on any design feature or consultant or underlying philosophy, on the quality of the staff. The CME (Center for Mathematics Education) Project team at Education Development Center (EDC) includes amazingly talented people: writers, teachers, mathematicians, and educators. In a sense, this project is more than a decade old, going back to *Connected Geometry* (Education Development Center 2000). It has been a wonderful experience to work with Nancy D'Amato, Jean Benson, Daniel Erman, Anna Baccaglini-Frank, Karen Graham, Brian Harvey, Wayne Harvey, Jud Hill, Bowen Kerins, Doreen Kilday, Helen Lebowitz, Steve Maurer, Melanie Palma, Sarah Sword, Audrey Ting, and Kevin Waterman.

But in addition to being too big, these courses are, at a deeper level, too small. There has been a growing consensus among all involved in secondary school mathematics education that this topic-driven curriculum is not serving our students well. More precisely:

1. The widespread utility and effectiveness of mathematics come not just from mastering specific skills, topics, and techniques but also, what is more important, from developing the ways of thinking—*the habits of mind*—used by scientists, mathematicians, engineers, and other professions in which mathematics is a core ingredient. Explicit emphasis on essential mathematical habits such as reasoning by continuity, abstracting regularity from repeated calculations, developing theories based on numerical evidence, and using thought-experiments are all but missing from many American programs (Cuoco 2006).

2. Mathematics as a scientific discipline is one of the crowning achievements of the human intellect. The RAND study panel (Ball 2003, p. 3) put it this way:

   > Mathematics constitutes one of humanity's most ancient and noble intellectual traditions. It is an enabling discipline for all of science and technology, providing powerful tools for analytical thought and the concepts and language for creating precise quantitative descriptions of the world. Even the most elementary mathematics involves knowledge and reasoning of extraordinary subtlety and beauty.

   But after decades of sustained and creative efforts, there is still a wide disconnect between mathematics as taught in many schools and mathematics as a scientific discipline. At the 2005 joint AMS-MAA meeting in Atlanta, talk after talk gave examples of how precollege courses emphasize low-level details, giving them the same importance as essential mathematical results, and how there is no overall mathematical point to many of the problems that students encounter. In far too many classrooms, from elementary school through undergraduate school, mathematics is taught as a disconnected set of facts and procedures, a body of knowledge to be learned in much the same way as one learns a list of terms for a vocabulary test.

Of course, this analysis is not new to the developers of the National Science Foundation (NSF)–funded curricula described in this volume. NSF has invested heavily in development efforts aimed at helping students become mathematical thinkers and at showcasing the immense utility and beauty of our subject.

The purpose of this chapter is to describe one of the newest such efforts, one that is similar in spirit and goals to the existing NSF-funded programs but that is based on a somewhat different philosophy and that uses a different set of organizing principles. Building on two prior curriculum efforts (Education Development Center 2000, 2001), my colleagues and I are developing a four-year high school program, the CME Project.

Some Design Features of the CME Project

Traditional Course Structure

The CME Project is a student-centered and problem-based program that adheres to the traditional American course structure: its courses have titles like Algebra, Geometry, and Precalculus. In addition to the logical reasons (described below) for such an organization, there is evidence, both from discussions with the teachers with whom we work and from surveys (e.g., St. John et al. 2000), that although high school teachers want to use problem-based materials and new methods of teaching, they are not as motivated to switch to integrated mathematics curricula or to unfamiliar organizations of material. So, the first reason for adhering to a conventional course structure is that there seems to be some demand in the field for it.

But another reason is that the ways in which ideas are organized in mathematics itself—essentially around the themes of algebra, geometry, and analysis—have emerged over the centuries as a scheme for bringing coherence to the discipline. Many standard curricula look at each of these areas as sets of results and techniques. Some integrated programs look at them as organizers that run through varying contexts. The CME Project team sees these branches of mathematics not only as compartments for certain kinds of results but also as descriptors for *methods* and *approaches*—the habits of mind that determine how knowledge is organized and generated within mathematics itself. As such, they deserve to be the core of a curriculum, not its by-products (Schmidt, Houang, and Cogan 2002). So, for example, an important ingredient in analytic thinking is *reasoning by continuity.* Starting in the geometry course and running through precalculus, we put students in situations where they develop a feel for the intermediate value theorem, build an understanding of contour lines for functions defined on geometric constructions, and develop the habit of looking at extreme cases in continuously changing systems. Some algebraic habits and their developmental trajectories are described later in this chapter.

The Role of Applications

"Power users" of mathematics are able to see abstract connections among seemingly different phenomena and can synthesize mathematical methods, often in unorthodox ways. And the first step in developing this proficiency is to expand students' conceptions of the "real world" to *include* mathematics.

This principle was crystallized for us several years ago when we invited some high school seniors to a meeting to discuss an earlier version of what will become our course for seniors (Education Development Center 2001). The meeting took place at the end of the fall term; up to that point, the students had been fitting polynomial functions to table values by looking at successive differences and other interpolation methods, and they had been proving that a recursively defined function and its closed form were equal on positive integers, using mathematical induction (Cuoco and Manes 2001). We asked them to tell us what was most different about this course. About four students (out of twelve) said, almost in unison, some variation of "It's more realistic." Others agreed. This reaction surprised even us, and as we "poked" at the remarks a bit, we discovered that the students meant that they were doing realistic *work*—the work they did in class and for homework went beyond the application of formulas; it asked them to reason their way around new situations, very much in the style of the intellectual work they did outside school. These students were typically not the "best" in the school—most of them did not have the grades to get into high school calculus—but they were motivated by the chance to use their own mathematical thinking.

Students in the CME Project *curriculum* apply elementary algebra and mathematical induction to determine the monthly payment on a loan; they use complex numbers as a device for establishing trigonometric identities; they use elementary arithmetic to study methods for creating secure ciphers; they apply Euclidean geometry to perspective drawing, optimization problems, and trigonometry. *All* these situations are applications of mathematics because the emphasis is on *how* one uses mathematics as opposed to *where* one uses it.

The Fundamental Dialectic: Open and Closed

The CME Project is the direct descendent of two previously developed courses, each using the traditional course structure and each focusing on mathematical thinking.

Connected Geometry (Education Development Center 2000) was developed immediately after the release of the 1989 *Curriculum and Evaluation Standards* (National Council of Teachers of Mathematics [NCTM] 1989), and it reflected many of the attitudes about curriculum that arose in that liberating period of American education. The book contained few stated theorems and definitions, and even fewer worked-out examples. The theorems, definitions, and examples were all there, but they were in the teacher's edition or the solution guide. What students saw was a collection of activities and provocative problems that were designed to help them discover the results for themselves.

Teachers who were not part of the original field tests told us that *Connected Geometry* was too much of a guide and not enough of a reference. They loved the open-ended problems, but they believed that the activities needed more closure—*in the student text.*

Mathematical Methods (Education Development Center 2001) was developed around the time that NCTM was revising the *Curriculum and Evaluation Standards* in preparation for *Principles and Standards for School Mathematics* (NCTM 2000). It, too, was a product of its time, influenced by the growing sense that all students needed a robust technical fluency (with algebraic and numerical calculations), and that students should be able to refer to their text as a resource for results and examples. *Mathematical Methods* contained many more proved theorems, worked-out examples, and "practice" exercises than we included in *Connected Geometry.*

With the CME Project, we have developed a design that is both shaped by our previous work and faithful to both needs: students can use their texts as both a guide and a reference. We have come to realize that at the high school level, understanding develops in two important ways: as the result of independent (or guided) investigations and as the result of reading, discussing, and internalizing mathematical exposition. Each CME Project activity starts out with a problem set that students do *before* instruction and that supplies experiments that preview—in simple numerical and geometric contexts—the important ideas in the exposition. The lesson then includes worked-out examples or written dialogues that codify methods, bring closure to this experimentation, and provide a reference for later work. In addition, each lesson has a set of *orchestrated practice problems* in which students practice arithmetic and algebraic skills while they try to abstract some regularity that suggests an interesting mathematical result.

Developing Mathematical Habits

Habits of mind are just that—habits that take time to develop. The CME Project's organization gives students the time and focus they need to develop central mathematical ways of thinking. But can beginning high school students really "think like mathematicians"?

We are convinced that they can. Although the problems of front-line research mathematics are out of reach for most nonspecialists, one of the wonderful things about our discipline is that the modes of thought used on the frontiers of what's known are natural extensions of ordinary human thought. And it turns out that developing this "mathematical mindedness" can go a long way toward helping students overcome what seem to be stubborn misconceptions about mathematics.

For example, the difficulties that students have with algebra word problems are legendary. A *mathematical* analysis of the difficulties students have in this area shows that the obstacles are related to the mathematical habit that's in play when one abstracts regularity from repeated calculations and compiles the actions into a coherent process, defining a mathematical function (Cuoco 1993). In our own high school teaching, we have exploited this common mathematical habit to develop a rather effective method (currently called "guess, check, and generalize") that helps students model situations with algebraic expressions.

Another example: The "Cartesian connection" between geometry and algebra is so ingrained in mathematical thinking that it's often a surprise to see how underdeveloped it is in algebra students, even among high school juniors and seniors. But we have seen repeatedly in high school classes that many students do not understand that one can test a point to see if it is on the graph of an equation by seeing if its coordinates satisfy the equation. Just *knowing* that this is a problem helps the CME Project writers design specific (and recurring) types of problems that address it. In the CME Project, we have developed a method (currently called the "point-tester method") that helps students derive equations for geometric objects by finding algebraic characterizations for the coordinates of the points on the objects.

For more on the habits of mind perspective, see Cuoco, Goldenberg, and Mark (1996). For more examples of curricular implications, see Cuoco (1998, 2005, 2006); Cuoco and Goldenberg (1997); and Cuoco and Levasseur (2003).

High Expectations

Reviews of early drafts of the CME Project chapters invariably contain comments like "High school kids could never do this kind of thing." Field tests show otherwise. The materials have to be revised, and often completely reworked, but in no instance do we need to water down the level of mathematics for either students or teachers. We are convinced that traditional curricula expect far too little from teachers and students and that students at all levels can do this kind of work. Much of the field testing of the materials for *Mathematical Methods* and for *Connected Geometry,* and now for the CME Project *Algebra 1* has taken place in "ordinary" classes. I cotaught two classes using *Mathematical Methods* materials for two years to students who were the weakest students in their school taking a fourth year of mathematics, and they were able to rise to my expectations. Poor performance in mathematics courses has many causes, but the lack of ability to think in a characteristically mathematical way is, for the vast majority of students, not one of them. The CME Project design employs a *low threshold–high ceiling* approach: each chapter starts with activities that are accessible to all students and ends with problems that will challenge the most advanced students. It's often (pleasantly) surprising to see how far students take the materials.

The Role of Technical Fluency

The CME Project team takes the position, coming from our teaching experience and work as mathematicians, that for the vast majority of students, the development of technical expertise in numerical and algebraic calculation is corequisite with the development of conceptual mathematical understanding. Of course, every teacher knows that the development of technical expertise does not guarantee mathematical understanding—we have all seen examples of students who can calculate like the wind and yet who cannot apply their skills to solve problems or develop theories.

But a great deal of algebraic thinking involves reasoning about calculations in algebraic structures, and in order to reason about calculations, students need to have some experience in performing them. And reasoning about calculations in abstract symbol systems is *useful.* For example, a manual for a commercial spreadsheet contains hundreds of examples, presented in a notation quite removed from traditional algebraic notation, that require this kind of algebraic thinking to understand. Finding one's way around the Internet, working through income tax software, even making flexible use of a word processor are all made easier by a knack for reasoning about operations and developing theories of calculations in self-contained systems.

In the CME Project, we invite students to become fluent in algebraic calculations so that they can reason about them. For example, we ask students to investigate the distribution of possible sums when three (or more) dice are thrown. We then ask them to find the coefficient of x^9 when

$$(x + x^2 + x^3 + x^4 + x^5 + x^6)^3$$

is expanded. The object here is *not* to perform the expansion by hand or machine but to calculate without calculating, reasoning where an x^9 term can occur when one multiplies out the expression.

When we have tried this with teachers and students, invariably someone says something like, "It's the number of ways you can make 9 as a sum of three numbers between 1 and 6—it's the number of ways you can roll a nine when three dice are thrown!" Such reasoning requires a kind of "decontextualization"—a formal approach to polynomial calculations that is important enough to deserve increased attention in the later years of high school.

Involving the Community

From the time we wrote *Connected Geometry,* we have taken the perspective that criticism from every corner of the mathematical community is essential to our work. We invite teachers, mathematics educators, and mathematicians to be on advisory boards, to consult with us, and to review the materials, and we take care to include people who are likely to have different points of view.

The CME Project has a teacher advisory board that met monthly during development cycles. It includes teachers who say they will never use a text that has worked-out examples and teachers who say that students should never tackle a problem unless they are given instruction on how to solve it. We can't follow either of these extremes, but this tension has had a substantial influence on our design, leading to the current structure of preliminary problem sets that are assigned with no instruction, the use of dialogues to convey exposition, the "for you to try" element that follows every worked-out example, and a host of other design elements.

Our national advisory board includes people with very diverse perspectives: Hy Bass, Dick Askey, Eric Robinson, Barbara Janson, Hung Hsi Wu, Herb Wilf, Glenn Stevens, Art Heinricher, Ed Barbeau, Arthur Eisenkraft, and Jackie Miller, among others. Roger Howe is an advisor as well as a core consultant. In addition to making for *very* spirited advisory board meetings, the varying perspectives we get from these advisors, consultants, and reviewers have greatly enriched our work and have made it highly likely that we hear about errors, criticisms, and points of view different from ours *before* the materials hit the street. As with our monthly meetings with teachers, we don't take all the advice we get, but even suggestions that are not implemented have an indirect effect on the finished product. In this way, we believe that the program reflects common wisdom across the entire community.

Developers of NSF-funded curricula know how valuable field testing is to the process. The CME Project is especially fortunate to work with a group of teachers who provide us with detailed feedback and criticism and who guide the development in ways that can come only from people who work every day with the materials.

Another important segment of the mathematics community is the publishing industry. Here, too, we think it is essential to incorporate the best advice we can get. Early in the development, we began talking to publishers with the express intention of establishing a genuine partnership that would complement our expertise. The CME Project will be published by Prentice Hall, and we have benefited greatly from our collaboration with the expert team that the publisher has devoted to this project.

Conclusion

Like any curriculum, the CME Project is not for everyone. The structure of the courses, emphasis on mathematical habits of mind, and approach to how mathematics is applied will appeal to some teachers and not to others. The development of specific mathematical topics (trigonometry, say) will resonate with some and grate on the nerves of others. Our taste in topics and focus is certainly not the only one. But our research and our own teaching experience convince us that there *is* a sizable audience for a program like the CME Project.

REFERENCES

Ball, Deborah Loewenberg. "Mathematical Proficiency for All Students: Toward a Strategic Research and Development Program in Mathematics Education." 2003. www.rand.org/pubs/monograph_reports/MR1643/index.html

Cuoco, Al. "Action to Process: Constructing Functions from Algebra Word Problems." *Intelligent Tutoring Media* 3/4, no. 4 (1993): 117–27.

———. "Mathematics as a Way of Looking at Things." In *High School Mathematics at Work.* Washington, D.C.: National Academy Press, 1998.

———. "Advanced Algebra in High School: Beyond Representation." In *Developing Students' Algebraic Reasoning Abilities,* Vol. 3, NCSM-Houghton Mifflin Company School Division and McDougal Littell Monograph Series for Leaders in Mathematics Education, edited by Carole Greenes and Carol Findell, pp. 46–60. Boston: Houghton Mifflin Co., 2005.

———. "Preparing for Calculus and Beyond: Some Curriculum Design Issues." In *A Fresh Start for Collegiate Mathematics: Rethinking the Courses below Calculus,* edited by Nancy Baxter Hastings, pp. 235–48. Washington, D.C.: Mathematical Association of America, 2006.

Cuoco, Al, and E. Paul Goldenberg. "Dynamic Geometry as a Bridge from Euclidean Geometry to Analysis." In *Geometry Turned On,* MAA Notes 41, edited by James R. King and Doris Schattschneider, pp. 33–44. Washington, D.C.: Mathematical Association of America, 1997.

Cuoco, Al, E. Paul Goldenberg, and June Mark. "Habits of Mind: An Organizing Principle for Mathematics Curriculum." *Journal of Mathematical Behavior* 15, no. 4 (1996): 375–402.

Cuoco, Al, and Ken Levasseur. "Classical Mathematics in the Age of CAS." In *Computer Algebra Systems in Secondary School Mathematics Education,* edited by James T. Fey, Al Cuoco, Carolyn Kieran, Lin McMullin, and Rose Mary Zbiek, pp. 97–116. Reston, Va.: National Council of Teachers of Mathematics, 2003.

Cuoco, Al, and Michelle Manes. "When Memory Fails: Putting Limitations to Good Use." *Mathematics Teacher* 94 (September 2001): 489–93.

Education Development Center. *Connected Geometry.* Chicago: Everyday Learning Corp., 2000.

———. *Mathematical Methods: Topics in Discrete and Precalculus Mathematics.* Armonk, N.Y.: It's About Time, 2001.

National Council of Teachers of Mathematics (NCTM). *Curriculum and Evaluation Standards for School Mathematics.* Reston, Va.: NCTM, 1989.

———. *Principles and Standards for School Mathematics.* Reston, Va.: NCTM, 2000.

St. John, Mark, Kasi A. Fuller, Nina Houghton, Dawn Huntwork, and Pamela Tambe. *High School Mathematics Curricular Decision-Making: A National Study of How Schools and Districts Select and Implement New Curricula.* Inverness, Calif.: Inverness Research Associates, 2000. www.inverness-research.org/reports/2000-01_Rpt_Compass_HSMathCurrDecisionMaking.pdf

Schmidt, William, Richard Houang, and Leland Cogan. "A Coherent Curriculum." *American Educator* (Summer 2002): 1–18.

10

The Case of

Core-Plus Mathematics

James T. Fey
Christian R. Hirsch

THE Core-Plus Mathematics Project was funded in 1992 by the National Science Foundation to develop a comprehensive *Standards*-based mathematics curriculum for grades 9–12. From the outset, our goal was to create a high school curriculum that would enable schools to successfully negotiate many of the difficult challenges of curricular reform outlined in *Everybody Counts* (National Research Council 1989), particularly

- expanding the traditional vision of school mathematics that offered minimal mathematics for the majority and advanced mathematics for a few in order to provide a significant core of important mathematics for all students;
- retaining the goal of offering a strong preparation for future studies in mathematics and its applications, but increasing attention to mathematical topics that are relevant to students' present and future needs.

The framework for the intended curriculum consists of a three-year core sequence of broadly useful mathematics for both college-bound and employment-bound students, plus a flexible fourth-year course that continues the preparation of students for college. Course 4 consists of a core of four units for all college-bound students, plus additional units enabling teachers to provide classes that support further mathematical preparation for either calculus-based or non-calculus-based undergraduate programs.

This chapter is based on work supported in part by the National Science Foundation (NSF) under grants no. MDR-9255257, ESI-9618193, and ESI-0137718. Any opinions, findings, and conclusions or recommendations expressed in this material are those of the authors and do not necessarily reflect the views of the NSF.

The co–principal investigators of the Core-Plus Mathematics Project are Christian Hirsch, Arthur Coxford (first edition), James Fey, and Harold Schoen. Collaborators on both the first and the second editions included Eric Hart (Maharishi University of Management), Brin Keller (Michigan State University), Beth Ritsema (Western Michigan University), and Ann Watkins (California State University, Northridge).

129

The team of mathematics educators working on the Core-Plus Mathematics program viewed the curriculum development process as an extended design experiment (Brown 1992; Collins 1992; Design-Based Research Collective 2003; Gravemeijer 1994) that includes cycles of curriculum material design, development, field testing, evaluation, and revision. We view professional development support for teachers and effective implementation of the curriculum to be an important part of our development responsibility.

The creation of any comprehensive school mathematics curriculum requires decisions that reflect the developers' understanding of mathematics as a discipline, their knowledge of research on learning and teaching, and their experience in schools. Advice and deliberation with those considerations led to design principles for the content and shape of the intended curriculum. The development of detailed materials reflecting that vision required imaginative and careful creation of plans and materials reflecting design principles.

Design Principles

The creation of the Core-Plus Mathematics curriculum and instructional materials has been guided by the development team's shared understanding of mathematics and effective approaches to teaching and learning. There are at least five dimensions of that vision.

Mathematical Content

The development of the Core-Plus Mathematics curriculum began with our belief that the essence of mathematics is its concepts and reasoning methods for making sense of observations and experiences in the real world. Concepts like similarity, equation, distribution, network, and algorithm help us to describe and analyze visual, numeric, and stochastic patterns. Operators like transformations, functions, logical inference rules, and matrices help us to extend patterns and solve problems. This view of mathematics as a broadly useful subject led us to several important design principles that shaped the scope and sequence of mathematical topics in the Core-Plus Mathematics curriculum:

- The curriculum should include major content strands that develop concepts and skills in algebra and functions, statistics and probability, geometry and trigonometry, and discrete mathematics.
- In addition to learning specific facts, principles, and procedures associated with the core content strands, the curriculum should pay explicit attention to developing students' mathematical habits of mind (Cuoco, Goldenberg, and Mark 1996) like visualizing, searching for and explaining patterns, thinking recursively, justifying and proving, and optimizing.
- The curriculum should explicitly develop students' understanding and skill in the use of mathematical modeling, including the processes of data collection, representation, interpretation, prediction, and simulation.

To the extent feasible within educational policy constraints of common school conditions, we have worked hard to follow these guidelines for the development of the mathematical content in the four-year curriculum.

Curriculum Sequence and Organization

Choosing the mathematical concepts and skills to be developed was an important first step in mapping our curriculum framework. But grade placement and ordering of the mathematical topics was not a trivial task. Our work on this problem has been guided by the following principles:

- The curriculum should include significant and broadly useful topics from each content strand in each year of the program.
- Topics from the separate content strands should be developed in coherent, focused units that exploit useful connections to the other strands.
- The mathematical content in any year of the curriculum should reflect judgments of what would be most important for students to know if that was to be their last formal experience in school mathematics.

Commitment to these curriculum sequence and organization principles has led us to a program that does indeed treat each main content strand in each year. Furthermore, we have departed from long-standing curricular traditions by placing topics where they seem most natural and important rather than where they have "always been." For instance, our development of formal algebraic symbol manipulation was gradually built up over all four years of the curriculum instead of being concentrated in the first and third high school years. More-complex manipulative skills were delayed to a point later in the curriculum when they could be studied intensively by students who would need them in future mathematical work. Our developments of formal logical inference and geometric proof occur somewhat later than tradition would dictate, although informal reasoning and the justification of results occur earlier and in more pervasive ways than in conventional U.S. curricula. Similarly, topics such as recursion, matrices, simulation (Monte Carlo) methods, and geometric transformations (including coordinate representations) occur earlier and in more pervasive ways than in more traditional curricula.

The Role of Technology

At the outset of our work on the Core-Plus Mathematics Project (CPMP), we were cautioned by teachers against making strong assumptions about access to technology like computers. Thus, we designed the first edition of CPMP curriculum materials with an assumption that students would have ready access only to graphing calculators. The fundamental aim in our use of technology was to enable multiple representations of mathematical ideas and to support a variety of robust strategies for mathematical thinking and quantitative problem solving.

Our first edition curriculum design and development did not assume ready access to emerging interactive geometry tools or computer algebra systems. However, in the revision of the Core-Plus Mathematics program, which is described later, we are making a more significant commitment to the use of interactive geometry software, spreadsheets, computer algebra systems, data analysis and simulation software, and tools for exploratory work in graph theory.

To some extent, our judgments about the reasonable impact of calculator and computer tools were influenced by our choice of the time frame for intended implementation. If our aim had been to develop curriculum ideas and materials that might be accepted for common school use by 2010 or later, we might have been more adventurous. However, our conscious goal in the development of the first edition of Core-Plus Mathematics was to produce curriculum materials that would help schools implement recommendations of the National Council of Teachers of Mathematics (NCTM) appearing in its *Curriculum and Evaluation Standards* (NCTM 1989) and later in its 2000 update described in *Principles and Standards for School Mathematics* (NCTM 2000).

Instructional Design

Traditional conceptions of mathematics education assign to textbooks the responsibility for clear and concise explanations of mathematical ideas and procedures, with illustrative examples to guide students' work on subsequent practice exercises. Of course, this model for curriculum materials is based on a conception of instruction that assigns the same role to teachers. Imaginative teachers have always enhanced textbook presentations by designing lessons that engaged students more actively in exploration and discovery of mathematical principles—believing that students learn and remember best those ideas that they have sorted out for themselves.

Design and development of the Core-Plus Mathematics curriculum materials were based on an explicit intention to support problem-based, student-centered classroom activity shaped by current theory and research on teaching and learning. As we developed materials for individual units and lessons, we aimed to implement the following instructional principles:

- School mathematics is best learned and understood as an active science of patterns involving quantity and change, shape and motion, data and chance, and enumeration and algorithms (Steen 1990).
- Any introduction of new mathematics will be most effective if the ideas and techniques appear in problem contexts that students can relate to and that connect to their prior knowledge. Authentic applied problems are especially useful contexts for learning, but significant pure mathematical problems are often useful also (Hiebert et al. 1996).
- Effective mathematics instruction frequently engages students in collaborative small-group investigations of problem situations that encourage student-to-student dialogue, followed by teacher-led whole-group summarizing activities that lead to analysis, abstraction, and further application of underlying mathematical ideas (Cobb 1995; Davidson and Kroll 1991). There is also some evidence that small-group collaborative learning encourages a range of social skills conducive to the learning styles of groups that are currently underrepresented in mathematics (Oakes 1990).

- Students should be regularly involved in mathematical activities like searching for patterns, making and verifying conjectures, generalizing, applying, proving, and reflecting on the process (Freudenthal 1983).

Commitment to these instructional design principles led us to produce curriculum materials that are intended to support a modified *launch–explore–summarize* instructional model. For each lesson, an introductory problem situation sets the context for the class investigation, and it is used for initial discussion involving the teacher and the whole class. Then students work in small groups to solve short sequences of selected problems and summarize their findings in response to several postinvestigation questions that are discussed with the teacher and the whole class. Homework tasks following the lessons are designed to engage students in applying, organizing, reflecting on, and extending their evolving mathematical understanding.

The instructional model for which Core-Plus Mathematics curriculum materials are written has led us to create student text material that is a thoughtfully constructed sequence of problems, not an archive of results and illustrative, worked examples. The investigations often involve hands-on experiments, data analysis and modeling, and technology-based exploration of patterns.

For example, an introductory lesson on the geometry of three-dimensional shapes begins with an experiment in which students test the load-bearing capacity of columns with constant perimeter but varying shapes. After collecting and organizing experimental data, students look for a numerical pattern relating the number of sides in the polygonal base shape to column load-bearing capacity. The lesson makes the important point that geometric form and function are closely related, and it also connects geometric properties to algebraic representations and numeric patterns.

Assessment Principles

Assessing what students know and are able to do is an integral part of the Core-Plus Mathematics curriculum's instructional model. There are opportunities for assessment in each phase of the instructional cycle. First, Think About This Situation questions in the lesson launch allow teachers to assess the prior knowledge that students bring to investigation of the new topic. While students work in small groups to explore new ideas and solve problems, teachers are able to monitor the group work and see where students are gaining insights or experiencing difficulties. An end-of-investigation Checkpoint provides explicit questions designed to guide the whole-class summary of investigation results and offers another opportunity to check student progress. Those questions are followed by On Your Own problems that give further feedback on students' understanding of the new material.

Since the student text is not a reference book of complete results (there are no answers in the student text for investigation or homework problems), students are prompted in several ways to construct their own summaries of important ideas and to record them in a mathematical toolkit journal that evolves over time. This record of students' thinking and understanding supplies them and their teachers with another useful tool to monitor the progress of their learning.

In addition to support for continual monitoring of students' progress during instruction, we also provide materials in print and electronic format for teachers to use or adapt as quizzes, unit examinations, project and take-home assessments, and mid-term and final assessments.

Development Process

The overall scheme of curriculum development in the Core-Plus Mathematics Project and other *Standards*-based reform projects of the past decade is similar in many respects to the iterative process outlined in the literature of design research (Design-Based Research Collective 2003), design experiments (Brown 1992; Collins 1992), developmental research (Gravemeijer 1994), and engineering research (Burkhardt and Schoenfeld 2003).

The Core-Plus Mathematics curriculum was developed in consultation with an international advisory board, mathematicians, instructional specialists with expertise in equity and access issues, and classroom teachers. Each course was the product of a four-year research, development, and evaluation process. After a year of initial development with local trials, the pilot version of each course was tested in nineteen Michigan high schools. During this pilot year, CPMP teachers in those schools provided extensive feedback to the authors by noting what worked and what was in need of revision. In addition, students in the CPMP classes were pretested at the beginning of Course 1 and posttested at the end of Course 1 and at the end of each course thereafter. The needed revisions identified by pilot-teachers' comments and test results were made promptly so that a revised, field-test version of a course was ready for use during the following school year.

The third year of development was the national field test, conducted in thirty-six high schools in Alaska, California, Colorado, Georgia, Idaho, Iowa, Kentucky, Michigan, Ohio, South Carolina, and Texas. A broad cross-section of students from urban, suburban, and rural communities with ethnic and cultural diversity was represented. Evaluative data, including evaluators' field notes, teacher-annotated units, and input from focus-group meetings with field-test teachers were used by the authors to make further revisions in the materials before they were finally published for wide use.

The orderly cycle of design, development, testing, evaluation, and revision described in theory is seldom matched in practice. In fact, our experience suggests that although the overall scheme follows the theory, there is almost continual interaction among all aspects of the process as a wide variety of opinions, advice, and external conditions are imposed on the development process.

The most carefully considered and widely agreed-on outline of a mathematical development for a unit or course often looks much less attractive when authors try to create problem material that will support the planned sequence of topics. Field-test focus groups often offer feedback on one unit that dictates changes that have serious implications for other units. Emerging state standards also introduce new considerations into the discussion of grade placement and topic coverage.

In the experience of developing the Core-Plus Mathematics curriculum, we found three factors that challenged smooth progress from vision to curriculum reality.

Field-Test Feedback

The Core-Plus Mathematics authors worked very hard to create engaging investigations that would lead students to discovery of important mathematics. They received detailed and thoughtful feedback from the authors of the other strands and the overall coordinating author and produced numerous iterations of draft material, even before field tests. However, despite those best efforts, feedback from the field-test teachers often doused authors' enthusiasm with a cold shower of reports that individual problems or whole investigations just didn't work. As discouraging as such reports sometimes are, they are essential to developing a curriculum that "works" in real schools.

This experience in the curriculum development process confirms two other important points that are often made about the design research process: "Development and research take place through continuous cycles of design, enactment, analysis, and redesign" (Design-Based Research Collective 2003, p. 5), and "Research must account for how designs function in authentic settings. It must not only document success or failure but also focus on interactions that refine our understanding of the learning issues involved" (Design-Based Research Collective 2003, p. 5).

In part, the often-discouraging reports from initial field tests reflect the reality that schools, classes, and teachers differ in very significant ways. Thus only quite robust curriculum designs and specific materials can be broadly useful without adaptation. However, they also highlight a particular challenge in writing materials that aim to support student-centered rather than teacher-directed instruction. We cannot say that we have found a way around this challenge of curriculum development, except to learn from feedback and to be creative in finding new approaches to the problematic topics.

Fidelity of Implementation

It is tempting to write off negative reports from field tests by guessing that the problematic material was simply not used as intended. In fact, there is a substantial challenge in fair testing of radical new ideas about classroom instruction and new mathematical goals for the curriculum. Field-test teachers often report puzzling over the level of mastery expected on topics in the new curriculum, since their experiential reference points have been knocked askew by the new content development. They also find it challenging to let students struggle a bit with open-end problem tasks. As a result, particularly in the first classroom testing of a new unit or course, it is unlikely that the material is taught as the authors envisioned.

In response to this challenge of implementing radical change and giving the new ideas a fair chance to succeed, we have made a variety of efforts to be sure we are reacting to reports that do indeed reflect experience with the intended curriculum. We monitor the field-test sites by visiting them as often as possible (not often enough), and we try to design field-test situations that give us a variety of contexts in which the material is being tested.

One of the most consistent reports from field tests is that there is more material in the curriculum than schools can reasonably teach. These reports can be tempered by other information (such as the sometimes astonishingly short class periods allowed for mathematics). But the reports of time pressure also make us alert to places where we have been too ambitious and where revision is required. Here again, the iterative cycle of writing, testing, and revision is essential to developing an effective curriculum package.

Pressures of Educational Policy Contexts

The field-test feedback and fidelity of implementation issues are directly related to, and somewhat controllable by, our own curriculum development process. However, our efforts are also affected in deep and often frustrating ways by the broader educational policy context for school mathematics. Our core ideas about curriculum content and organization, about classroom instruction, and about assessment of students' learning all challenge long-standing traditions in U.S. school mathematics. Despite the consistent judgment that those traditional practices are not notably successful, all proposals for radical change are met with spirited resistance both within and outside the mathematics education community. For many teachers, challenges to change long-standing standards of practice are often heard as criticisms of their instructional skills. For parents and policymakers, challenges to change curriculum goals suggest risks of failure in the pervasive network of high-stakes tests used to measure schools' progress and students' qualification for advancement and admission to higher education.

For the Core-Plus Mathematics Project, challenges to our ideas have led to three basic kinds of response. First, we have tried to allay fears about failure on high-stakes tests by collecting extensive evaluation data (cf. Schoen and Hirsch 2003) that can offer reassurance (where it is positive or at least neutral in comparison with traditional testing). Sometimes those data are empirical results from focused studies; sometimes they are simple reports from college admission officers who generally suggest that participation in a reform mathematics program will not be a barrier to higher education. Second, we have found that successful Core-Plus Mathematics teachers and school systems are often the most effective voices for communicating potential benefits of innovation to skeptical schools and teachers. As a result, we have worked closely with districts to develop local evaluation plans and to disseminate case study reports of local experiences (cf. www.wmich.edu/cpmp/districtreports.html).

Third, we have provided a variety of supplementary materials to "cover" particularly problematic topics that high-stakes tests still honor but are not high on our list of curriculum priorities. The current revision of the Core-Plus Mathematics curriculum is also responding to persistent pressures from the field (reflected in states' responses to No Child Left Behind legislation and college placement testing) by modifying the course placement of important topics (generally, formal manipulative algebra skills).

Design Principles Revisited

In 2002, the Core-Plus Mathematics Project received a five-year award from the National Science Foundation to prepare a revision of the Core-Plus Mathematics curriculum. The revision is being shaped by research on the program's effectiveness, including a five-year longitudinal study, and by extensive feedback from teachers using the first edition texts. The revision is also taking into account recent changes in middle school mathematics programs (particularly in the area of algebra), the evolving nature of undergraduate mathematics, and advances in technology.

Content Sequence and Organization

The mathematical content and organization of the revised Core-Plus Mathematics curriculum reflects the project's continuing commitment to preparing students well for "the needs of all disciplines and careers in which mathematical tools are used," by offering "a coherent, balanced introduction to the most widely used parts of the mathematical sciences in a manner that regularly connects each part with several others" (Steen 2007, p. 92). The revision work, however, has resulted in some shifts in the positioning and priorities of mathematical content in the four-year curriculum, most notably in the algebra and functions and the geometry and trigonometry strands. In the algebra and functions strand, changes resulted in

- an accelerated introduction of symbol-based reasoning—including symbol manipulation and proof;
- more explicit development of symbol sense—connecting algebraic forms with numeric, graphic, and context interpretation and implications;
- earlier introduction of inverse functions and logarithms.

In the geometry and trigonometry strand, major content changes resulted in

- a reorganized development of important geometric ideas—congruence, properties of circles, and trigonometric ratios and functions;
- earlier and more explicit development of geometric reasoning and proof;
- increased use of coordinate representations and algebra to support geometric reasoning and problem solving.

Instructional Design

The instructional materials themselves reflect the same design principles that guided the development of the first edition, albeit with some refinements based on reports from users of the first edition. Refinements include streamlined lessons with focusing question(s) at the beginning of each investigation to provide an advance organizer for the mathematics to be discovered or developed. For example, in *Core-Plus Mathematics, Course 1* (Hirsch et al. 2008), the introduction to an investigation in the second lesson of the "Quadratic Functions" unit includes the following focusing questions:

As you work on the problems in this investigation, look for answers to these questions:

What strategies are useful in finding rules for quadratic functions?
In deciding when two quadratic expressions are equivalent?
In deciding when one type of quadratic expression
is more useful than another?

A second refinement is the inclusion of Review exercises in each homework set to build proficiency with concepts and skills through distributed practice, to provide just-in-time review of background for coming investigations, or to address possible gaps in prerequisite preparation caused by our more robust assumptions about middle school preparation of students.

The Role of Technology

Because of concerns for access and equity, the first edition's curriculum materials were based on a modest technology assumption—students would have access to graphing calculators in class and outside of school. As work began on the second edition, both the contextual and mathematical problems around which the curriculum was being organized and the learning expectations for students were such that it was desirable to augment graphing calculator technology with computer tools. To meet this challenge and maintain the project's commitment to access and equity, the project systematically explored development of Java-based software that eventually evolved into *CPMP-Tools*—a suite of tools whose development continues to be shaped by, and integrated with, the development of the curriculum materials.

• Tools were developed for each strand of the curriculum—algebra, geometry, statistics, and discrete mathematics.

Algebra—The software for work on algebra problems includes an electronic spreadsheet and a computer algebra system (CAS) that produces tables and graphs of functions, manipulates algebraic expressions, and solves equations and inequalities.

Geometry—The software for work on geometry problems includes an interactive drawing program for constructing, measuring, manipulating, and transforming geometric figures and a set of custom tools for studying geometric models of physical mechanisms, tessellations, and special shapes.

Statistics—The software for work on data analysis and probability problems provides tools for graphic display and analysis of data, simulation of probability experiments, and mathematical modeling of quantitative relationships.

Discrete Mathematics—The software for work on graph-theory problems enables students to construct, manipulate, and analyze vertex-edge graphs and networks.

• Developing students' disposition and ability to make decisions about what technology tool to use and when was an important consideration. The design of *CPMP-Tools* keeps the possibilities up front.

- Tools are built using Java WebStart, which permits safe, easy, and reliable distribution of software and software updates across different types of computers.

From the outset, our goal was to develop learner-centered software built and available through an open-source license. Given the unknown future, an open-source license will help to ensure that others could both maintain and build on our work. Furthermore, by using Java WebStart, we have situated the tools in a manner making them easily upgraded in the future. Such a delivery mechanism also supports the potential use of the software in libraries, at home, or any location where students have Internet access.

Further information about the design and capabilities of this evolving curriculum-embedded software can be found in Hart, Hirsch, and Keller (2007).

Teacher Support

Close work with a small group of teachers during the development of the second edition, along with information from the field during the last ten years and reviews of the first edition's teacher-support materials have prompted enhancements to these materials. These enhancements are designed to better support teachers' learning and effective implementation of the curriculum.

The enhancements have focused on more clearly conveying the mathematical goals at the unit, lesson, and sometimes problem level. This includes helping teachers understand the level of students' proficiency expected at different points on the learning trajectories of important ideas. In addition, enhancements have been developed to help teachers better understand how students might approach problems within the investigations. This has been done through notes labeled "Common Error," "Misconception," or more generally, "Instructional Note." In addition, newly developed "Differentiation" notes support teachers in making curriculum and instructional modifications to meet the needs of all students.

In addition to continuing to support teachers' learning of mathematics, we are exploring ways to make the teacher-support materials more educative about pedagogical practice. We are drawing on a design framework developed by Davis and Krajcik (2005). For example, to help teachers think more deeply about the lesson launches and the investigation summaries, we have developed discourse scenarios to provide annotated examples of student and teacher discourse at these pivotal stages of a lesson. These scenarios have been written in such a manner that teachers should be able to "see" their own students in the discussions. Embedded in these scenarios are follow-up questions that teachers might choose to use during the discussion along with occasional parenthetical remarks explicating possible moves by the teacher.

By carefully examining the student masters produced by field-test teachers, we gained insights into teachers' decisions to scaffold certain investigations. This information along with teachers' notes in annotated field-test units provided guidelines for developing some additional student activity masters for the second edition that assist students in organizing their work without reducing the cognitive demand of the mathematical tasks or redirecting the goals of the investigation.

The Development of the Second Edition

The curriculum development process for the second edition remains similar to that used for the first edition of the Core-Plus Mathematics curriculum. Each revised text is the product of a cycle of (1) research and development, (2) pilot testing and refinement, (3) field testing in an expanded set of schools, and (4) further refinement prior to publication. In addition to the careful reviews by teachers and the project's advisory board, work on the revision project was critiqued by a special panel of mathematical consultants. This panel reviewed and commented on units as they were being developed, tested, and refined. This modification of our original development process has worked well and has proved to be beneficial and rewarding for both developers and consultants (cf. Maurer and McCallum 2006).

Summary

The conceptualization, development, and evaluation of the Core-Plus Mathematics curriculum materials has been a collaborative effort that blends the mathematical and pedagogical insights and practical experiences of mathematicians, mathematics curriculum and assessment developers, and classroom teachers. We believe that this collaboration offers a foundation of sound mathematical judgments, access to advice from the best of current thinking and research about learning and teaching, and the wisdom of practice that makes it possible for strong and appropriate new curriculum ideas to work in real school situations.

Although there is contentious public discourse about school mathematics that suggests a disconnect of the prominent players in curriculum work, we have found that it is feasible and desirable to draw mathematicians, mathematics education researchers, and classroom teachers together in productive collaboration. Such collaboration requires sufficient time for extended planning, for dialogue about content and teaching issues, and for a development process that involves iterative cycles of design and field trials through which the expertise of all parties can be applied to the task of producing excellent materials.

The use of curriculum materials in diverse classroom settings is a complex process that is shaped by the influence of many different ideas, institutional traditions, and personal values. Thus it is not surprising that our own experience in developing the Core-Plus Mathematics program has been far from a smooth unfolding of designs that are logical consequences of mathematical, pedagogical, and design-process principles. However, we believe that our operation under the umbrella of a generally shared set of such process principles has allowed us to produce a coherent and effective new approach to the mathematical education of high school students.

REFERENCES

Brown, Ann L. "Design Experiments: Theoretical and Methodological Challenges in Creating Complex Interventions in Classroom Settings." *Journal of the Learning Sciences* 2, no. 2 (1992): 141–78.

Burkhardt, Hugh, and Alan H. Schoenfeld. "Improving Educational Research: Toward a More Useful, More Influential, and Better-Funded Enterprise." *Educational Researcher* 32, no. 9 (2003): 3–14.

Cobb, Paul. "Where Is the Mind? Constructivist and Sociocultural Perspectives on Mathematical Development." *Educational Researcher* 23, no. 7 (1995): 13–20.

Collins, Allan. "Toward a Design Science of Education." In *New Directions in Educational Technology,* edited by Eileen Scanlon and Tim O'Shea, pp. 15–22. New York: Springer-Verlag, 1992.

Cuoco, Al, E. Paul Goldenberg, and June Mark. "Habits of Mind: An Organizing Principle for Mathematics Curricula." *Journal of Mathematical Behavior* 15 (1996): 375–402.

Davidson, Neil, and Diana Lambdin Kroll. "An Overview of Research on Cooperative Learning Related to Mathematics." *Journal for Research in Mathematics Education* 22 (November 1991): 362–65.

Davis, Elizabeth A., and Joseph S. Krajcik. "Designing Educative Curriculum Materials to Promote Teacher Learning." *Educational Researcher* 34, no. 3 (2005): 3–14.

Design-Based Research Collective. "Design-Based Research: An Emerging Paradigm for Educational Inquiry." *Educational Researcher* 32, no. 1 (2003): 5–8.

Freudenthal, Hans. *Didactical Phenomenology of Mathematical Structures.* Dordrecht, Netherlands: D. Reidel Publishing Co., 1983.

Gravemeijer, Koeno. "Educational Development and Developmental Research in Mathematics Education." *Journal for Research in Mathematics Education* 25 (November 1994): 443–71.

Hart, Eric W., Christian R. Hirsch, and Sabrina A. Keller. "Amplifying Student Learning in Mathematics Using Curriculum-Embedded Java-Based Software." In *The Learning of Mathematics,* Sixty-ninth Yearbook of the National Council of Teachers of Mathematics (NCTM), edited by W. Gary Martin and Marilyn E. Strutchens, pp. 175–204. Reston, Va.: NCTM, 2007.

Hiebert, James, Thomas P. Carpenter, Elizabeth Fennema, Karen Fuson, Piet Human, Hanlie Murray, Alwyn Olivier, and Diana Wearne. "Problem Solving as a Basis for Reform in Curriculum and Instruction: The Case of Mathematics." *Educational Researcher* 25, no. 4 (1996): 12–21.

Hirsch, Christian R., James T. Fey, Eric W. Hart, Harold L. Schoen, and Ann E. Watkins, with Beth Ritsema, Rebecca Walker, Sabrina Keller, Robin Marcus, Arthur F. Coxford, and Gail Burrill. *Core-Plus Mathematics, Course 1.* Columbus, Ohio: Glencoe/McGraw-Hill, 2008.

Maurer, Stephen B., and William McCallum. "Advising a Precollege Curriculum Project." *Notices of the AMS* 53, no. 9 (2006): 1018–20.

National Council of Teachers of Mathematics (NCTM). *Curriculum and Evaluation Standards for School Mathematics.* Reston, Va.: NCTM, 1989.

———. *Principles and Standards for School Mathematics.* Reston, Va.: NCTM, 2000.

National Research Council. *Everybody Counts: A Report to the Nation on the Future of Mathematics Education.* Washington, D.C.: National Academy Press, 1989.

Oakes, Jeannie. "Opportunities, Achievement, and Choice: Women and Minority Students in Science and Mathematics." In *Review of Research in Education,* Vol. 16, edited by Courtney B. Cozden, pp. 153–222. Washington, D.C.: American Education Research Association, 1990.

Schoen, Harold L., and Christian R. Hirsch. "The Core-Plus Mathematics Project: Perspectives and Student Achievement." In *Standards-Based School Mathematics Curricula: What Are They? What Do Students Learn?* edited by Sharon L. Senk and Denisse R. Thompson, pp. 311–43. Mahwah, N.J.: Lawrence Erlbaum Associates, 2003.

Steen, Lynn Arthur. "Facing Facts: Achieving Balance in High School Mathematics." *Mathematics Teacher* 100 (January 2007): 86–95.

Steen, Lynn Arthur, ed. *On the Shoulders of Giants: New Approaches to Numeracy.* Washington, D.C.: National Academy Press, 1990.

11

The Case of the

Interactive Mathematics Program

Sherry Fraser

I N 1988, the state of California, under the leadership of Walter Denham, issued a request for proposals for totally revamping the traditional sequence of high school mathematics. In place of Algebra 1–Geometry–Algebra 2, the proposal guidelines called for a modern mathematics curriculum with problem solving, reasoning, and communication as major goals. The new curriculum would also include such areas as statistics and discrete mathematics and make use of the latest technology (California Postsecondary Education Commission 1988). Additionally, the new curriculum had to be flexible enough to meet the needs of all college-bound students and lower the attrition rate of students in the col-lege-prep sequence, especially women and minority students. Funding for the development and implementation of the Interactive Mathematics Program (IMP) started with public funds from the California Postsecondary Commission, the U.S. Department of Education, and the National Science Foundation. Money and support also came from private foundations such as the Noyce Foundation, the David and Lucile Packard Foundation, the San Francisco Foundation, the Stuart Foundation, and the Intel Foundation.

A comprehensive, four-year program of problem-based mathematics was pub-lished after more than ten years of research, pilot tests, evaluations, field tests, revisions, and detailed reviews by professionals in the field. The IMP curriculum is used by thousands of teachers across the United States. It has been translated into Spanish, French, Korean, Hawaiian, and Chinese. How did IMP grow from an idea in 1988 to today's comprehensive program, which includes regional support centers and professional development opportunities for teachers, teacher-leaders, and directors of regional centers? It all started with a basic set of principles around curriculum, instruction, assessment, equity, and teacher support.

The Interactive Mathematics Program was developed by Diane Resek and Dan Fendel (San Francisco State University) and Lynne Alper and Sherry Fraser (Lawrence Hall of Science, University of California, Berkeley). Key Curriculum Press is the publisher.

Program Design

Curriculum

Meaningful mathematics

The IMP curriculum focuses on practical mathematics and facilitates an understanding of abstractions by providing concrete experiences. This curriculum approach is described by the mathematician and education researcher Hans Freudenthal as "Realistic Mathematics Education." Freudenthal found that the practical investigation of mathematics provides students with opportunities to develop their own ways of thinking and mathematizing, and with the ability to see how mathematics can be applied to real situations (Gravemeijer 1994, p. 445). The approach meets a broad range of students' needs and addresses individual learning styles.

In the IMP curriculum we wanted to introduce abstract ideas as concretely as possible, so that the mathematics becomes accessible to fourteen-year-olds. Traditional mathematics curricula usually start a new topic by presenting an abstract generalization to students and then ask the student to apply the generalization to specific cases. However, most people learn more effectively by starting with practical, realistic situations (Turkel and Papert 1992). Learners can proceed by getting involved in the details of the problem and then make generalizations based on their experiences.

Problem-solving focus

Studying mathematics in the context of problems motivates students to think mathematically and to make connections between skills and concepts from all mathematical areas: algebra, geometry, trigonometry, statistics, and probability. Curriculum materials in which students first engage in a design project or in large-scale problem solving encourage students to use the informal understandings and experiences they bring to school so that the learning of mathematics can build from those understandings (Kilpatrick, Martin, and Schifter 2003). "This will be on the test" or "you will need to know this next year" motivates some students, but most students need more intrinsic motivation to work on mathematics. High school educators need to reach all students, not just those who accept the idea that mathematics is part of their future.

Content strands

The original funding from the California Postsecondary Education Commission required the content of the curriculum to match the mathematical expectations for freshmen entering the University of California. The National Council of Teachers of Mathematics (NCTM) *Curriculum and Evaluation Standards for School Mathematics* (NCTM 1989) had just been published. Using these two documents as guides, the IMP directors organized the recommended mathematics by strands as recommended in the NCTM *Standards*. But rather than teach each strand in isolation, it was decided to develop each of the major strands in an integrated way. The goal was to help students develop an in-depth, coherent understanding of mathematical concepts and techniques and their applications over a four-year period. So the first question the IMP curriculum developers had to answer was: How do you introduce all the strands in the ninth-grade year, and then come back and build on the learning experiences each year, in each of the strands?

Unit design

We decided to organize the content into six- to eight-week units. Each unit would begin with a central problem or theme. Students would then explore and solve that problem over the course of the unit. Each unit had to be more than a collection of lessons and activities; the mathematical work must be coherent. The work students do must include a wide variety of tasks, all clearly related to the primary goal, that lead toward clarification and consolidation of a set of general ideas that will be useful later. Each unit must interweave strands, ensuring breadth, and also deepen one or more unifying ideas.

Units should develop all the dimensions of mathematical power, support students working collaboratively and independently, develop students' positive dispositions toward mathematics, and take into account historical, societal, and career information. In addition, assessment should be integrated with instruction in the unit (California Department of Education 1992). As the units were developed, the following questions were used as a guide to the suitability of a unit:

- Does the subject matter represent an important area of knowledge?
- Does the mathematics arise naturally in the unit?
- Are mathematical ideas integral to the understanding of the unit?
- Does the unit have natural coherence?
- Does the mathematics in the unit have important internal connections within mathematics?
- Does the unit provide external connections to work in other disciplines?
- Is the appropriate technology used to enhance learning?

Communication skills

A mathematics classroom in which teachers emphasize communication and language development helps students build the capacity to think, reason, solve complex problems, and communicate. The IMP curriculum insists that students describe their ideas both orally and in writing, develop conjectures based on their own investigations, and explain how they arrived at their solutions or conclusions. Rather than give students prepackaged methods, the assignments actively encourage students to make sense of the mathematics and develop procedures that evolve from their thinking. By encouraging and acknowledging students' varied methods of solution, teachers convey to students that their thinking is valued. When a student explains his or her approach during class discussion, other students feel comfortable exploring future problems without feeling they have to memorize the "right" approach.

Technology

Current technology has changed both how mathematical problems are solved outside of school and the mathematics content students need to solve problems in the real world. Graphing calculators allow students to focus on the mathematics in a problem without getting bogged down with computation. Computers can be used to solve problems and accomplish tasks that would otherwise be very difficult or impossible. Therefore, the IMP curriculum developers decided to make in-class activities graphing calculator–dependent, to make homework calculator-dependent, and to use computer technology whenever it was feasible.

Instruction

Active learning

In many traditional classrooms, a student's task is to mimic the work presented by the teacher and to find numerical answers to similar problems. But in a world that is ever changing, students need to be equipped to handle problems they have never seen before, and to handle them with confidence and perseverance. The IMP curriculum is designed to give students a more active part in their learning. They work with complex and realistic situations rather than problems fitting a rigid format. They construct new ideas by moving from specific examples to general principles. They progress beyond simply finding numerical answers; they use those answers to make decisions about real-life problem situations. Because the curriculum moves beyond mechanical skills, the teacher's role must expand as well. The teacher asks challenging questions and provokes students to do their own thinking, to make generalizations, and to discern connections and relationships.

Collaborative learning

Collaborative work in the classroom can have positive effects on students' achievement and can contribute to students' productive disposition toward mathematics. Both formal research and teachers' observations indicate that active classrooms maintain the interest of many students who do not do well in traditional, more passive learning situations (Kilpatrick, Swafford, and Findell 2001). By communicating their ideas to others, students reach deeper levels of understanding. At the same time that students are expanding their ability to work together, they are gaining independence as learners and thinkers.

Embedded Assessment

It is essential to identify and use effective assessment techniques that enhance thinking in classrooms and that give more complete information for teaching decisions and for external evaluation purposes. Assessment in the classroom is an ongoing, daily process and takes many forms, including daily homework assignments, oral presentations, contributions to the group or whole-class discussion, students' self-assessments, and students' portfolios.

Commitment to Equity

One of the great strengths of IMP is its commitment to diversity and equity. The original project included two mathematicians from San Francisco State University, two teacher educators from the University of California, Berkeley, and six teachers from three different high schools: an inner-city school, an urban school, and a rural school. Each of the four directors had previous experience in curriculum development and professional development. Prior to working on IMP, both mathematicians wrote college-level textbooks and designed courses for underachieving students. They also worked with the grades K–12 school community, teaching in the schools and designing mathematics and computer curricula. The two teacher educators had many years of practical experience teaching mathematics in high schools and had planned, organized, and conducted mathematics in-service programs for secondary school teachers for many years, with special focus on access and equity. In addition, both teacher educators were experienced curriculum developers and had published grades K–12 curriculum books for mathematics education and computer education.

The original six IMP pilot teachers also brought a wealth of experience. The teachers from the rural school had just spent a year looking for integrated mathematics curriculum that would support their International Baccalaureate program. The teachers from the inner-city school had developed a new course for their students. It was designed to prepare them to take challenging college preparatory mathematics classes. At the large urban school, teachers knew they were leaving many minority students behind by a tracking system that placed students in dead-end mathematics classes.

The expertise of all ten educators was needed to make the IMP vision a reality. But just as important were the students. One hundred students were selected at each of the three high schools to be part of the initial curriculum development process. These heterogeneous groups of students were part of the process, and they kept everyone honest. It was real curriculum development in real time, every day, for three years.

Teacher Support

Professional development

To implement the IMP curriculum, mathematics teachers are asked to make major changes in what they teach and how they teach. Appropriate support and training are crucial elements in teachers' success as they work to improve their teaching. Therefore, designing the professional development for teachers was just as important as developing the curriculum. To support the teachers and their schools, a comprehensive, ongoing professional development program was created.

Professional community

As the number of schools using the IMP curriculum expanded throughout the country, a network of regional centers was established to empower others to provide support for teachers. Teachers prefer support at the local level with educators who are as committed as they are to improving mathematics education. This network of IMP regional directors has been meeting for eleven years for their own professional development, as well as guiding and supporting curriculum implementation in their regions.

Program Development

Process

The core team tried out and evaluated almost every activity themselves and with teachers at workshops before ever using the activities in the classroom. As units were being developed, subject-matter experts reviewed initial drafts before field testing. But the real test was in the classroom with the 100 students at each of the three pilot-test schools. In addition to the six pilot-test teachers, two of the IMP directors team-taught the curriculum at one of the pilot-test schools. The other two directors sat in classrooms every week, talking with students and observing their reactions to the activities. The pilot-test teachers were given an extra professional development period each day so they could discuss the successes and challenges of the instructional tasks, and the unit as a whole, before meeting with the entire group. Every six weeks the directors and pilot-test teachers met for approximately two days to give feedback on what was happening in the classroom and to prepare for the next unit that they would be teaching. Winter retreats and summer institutes were also held.

The original intent was to focus on only three schools to develop the curriculum and provide an existence proof of another way to teach high school mathematics. But by the end of the first year, colleagues at other schools wanted to be part of the development and some had funds to support teachers with ongoing extra professional development periods. Supplemental funding was sought from the National Science Foundation. The project expanded from three schools the first year, to six schools the second year, to twelve schools the third year. This larger group of fifty teachers became known as the Teacher Advisory Group. These new schools represented suburban schools and urban schools from four other states, and the teachers involved at these schools provided valuable feedback. Each unit was revised after its first pilot testing, and the revised units were tested again in the new schools and with the second cohort of students at the three original schools.

In 1992, IMP received a five-year grant from the National Science Foundation to write a fourth-year curriculum, strengthen the curriculum for the first three years by rewriting the units, evaluate the effectiveness of the complete curriculum for students, assess the needs for teacher in-service training, and disseminate the program widely through regional centers as a focus of implementation. With this additional funding, every IMP unit underwent another thorough reexamination with a team of experienced classroom teachers with longitudinal experience from the original extensive field test. Four to six teachers who had already taught field-test units were funded to keep detailed notes as they taught the units again. They then met with the IMP directors for one or two days for a reexamination of each unit. This led to yet another revision and further improvements.

In 1994, the entire revised IMP program was ready for one final field test as a four-year sequence. The three new schools selected for the final field test had not been involved in the original curriculum development. From 1994 to 1998, one of the IMP directors team-taught the curriculum at one of the schools. Each teacher kept detailed notes of what happened in the classroom. At the end of each unit, the teachers met with the IMP directors to discuss the unit and to identify potential trouble spots and share possible solutions. The directors revised the units again before handing them over to Key Curriculum Press to publish. The curriculum was published one year at a time, with Year 1 available in fall 1996 and Year 4 available in fall 1999.

Potential Barriers

One potential barrier to implementing the program was the belief by some administrators, counselors, teachers, students, and parents that one curriculum could not serve students of varied backgrounds in heterogeneous classrooms. Our experience with IMP shows that a curriculum built around complex, open-ended problems can be explored at many levels of sophistication. The central problems in the units have a richness that challenges the brightest students and a concreteness that allows all students to do meaningful work. In addition, supplemental problems extend or reinforce the content.

Another potential barrier was to find additional time for teachers to reflect on and to discuss their role as a teacher. A teacher using the IMP curriculum must be a keen observer, an active listener, and a skilled facilitator to ensure that students progress in their learning. The teacher needs to ask challenging questions and provoke students to do their own thinking, to make generalizations, and to discern connections and relationships among mathematical ideas. All this takes time, reflection, and practice. Therefore, before the curriculum was published in 1996, all teachers who wanted to pilot-test, field-test, or use the draft curriculum materials were required to have an extra planning period. This would not have been possible without the original California grant, support from private foundations, and the large National Science Foundation grant. These monies allowed the project to overcome this barrier and implement the curriculum as designed.

One barrier that was not overcome was successful integration of computer technology into the curriculum. During the pilot-test stage, the directors found that many high schools did not have access to computers for math students. Trying to schedule computer labs for all the mathematics classes that needed them became a logistical nightmare. However, almost all teachers could get one computer in their classroom, so a decision was made to design the curriculum for this type of situation and use classroom sets of graphing calculators as the primary source of technology.

Because of extensive support over a long period of time, the project was not required to compromise on any of its core beliefs. We were given the time and the money to carry out our vision. The directors met once a week for more than eight years and took advantage of their diverse expertise to develop a curriculum that better serves all students.

Lessons Learned and Implications for the Future

Thoughtful design principles alone, even when accompanied by an adequate development process, are not enough to ensure the successful implementation of a new curriculum. Other factors essential to the long-term use of a curriculum are the diversity and consistency of the leadership, the commitment to equity, the integration of professional development with the curriculum, the establishment of a network of regional centers to support the implementation, and a supportive publisher who shares your vision.

Although this was a curriculum development effort, the IMP directors proposed that they create the professional development to accompany the curriculum and establish regional centers around the country as part of the grant. The IMP regional centers and the ongoing professional development of teachers have kept the IMP program strong. Regional center directors are asked to assume the role of leader of leaders and commit to building a leadership team among the teachers. The directors teach each year of the curriculum and model peer coaching. They work with schools and districts to disseminate information and materials to teachers, curriculum coordinators, and administrators. The directors at the centers are also responsible for community outreach. They create and maintain informational packets to provide to visitors and to feeder school districts. They coordinate the recruitment of new students at feeder schools by presenting IMP to eighth-grade mathematics teachers and parent groups. They also coordinate the professional development and public relations for their region.

In conclusion, curriculum is a vehicle to create change in the mathematics classroom. It is one piece of the effort to improve mathematics education. Alone, it is not enough.

REFERENCES

California Department of Education. *Mathematics Framework for California Public Schools.* Sacramento, Calif.: California Department of Education, 1992.

California Postsecondary Education Commission. *Request for Proposals, Secondary Mathematics Eisenhower Grant.* Sacramento, Calif.: California Postsecondary Education Commission, 1988.

Gravemeijer, Koeno. "Educational Development and Developmental Research in Mathematics Education." *Journal for Research in Mathematics Education* 25 (November 1994): 443–71.

Kilpatrick, Jeremy, Jane Swafford, and Bradford Findell, eds. *Adding It Up: Helping Children Learn Mathematics.* Center for Education, Division of Behavioral and Social Sciences and Education. Washington, D.C.: National Academy Press, 2001.

Kilpatrick, Jeremy, W. Gary Martin, and Deborah Schifter, eds. *A Research Companion to "Principles and Standards for School Mathematics."* Reston, Va.: National Council of Teachers of Mathematics, 2003.

National Council of Teachers of Mathematics (NCTM). *Curriculum and Evaluation Standards for School Mathematics.* Reston, Va.: NCTM, 1989.

Turkel, Sherry, and Seymour Papert. "Epistemological Pluralism and the Revaluation of the Concrete." *Journal of Mathematical Behavior* 11, no. 1 (1992): 3–34.

12

The Case of

MATH *Connections*®

William P. Berlinghoff

MATH *Connections* began in 1992 with a five-year National Science Foundation grant awarded to the Education Foundation of the Connecticut Business and Industry Association (CBIA),[1] whose interest in mathematics education stemmed from the inadequate mathematical skills of their members' potential employees. As stated in the grant proposal, the overarching mission of the project was the development of a comprehensive three-year, core curriculum designed to serve the needs of *all* students and to make mathematics more appealing and more accessible to them. It was envisioned that this common, three-year core would be followed by a choice of fourth-year options tailored to students' differing educational and career goals, preparing them for postsecondary education or the workplace. A distinctive feature of the proposal was the planned use of the corporate resources of the member companies of the CBIA in developing the curriculum and in providing authentic applications from business and industry.

To implement this plan, several teams of people with expertise and interest in playing complementary roles were assembled.

The **MATH** *Connections* curriculum was developed by the Education Foundation of the Connecticut Business and Industry Association (CBIA). The principal investigators were project director June G. Ellis (CBIA Education Foundation), Robert Rosenbaum (Wesleyan University), and Robert Decker (University of Hartford). William Berlinghoff (Colby College), Clifford Sloyer (University of Delaware), Robert Hayden (Plymouth State College), and Eric Wood (University of Western Ontario) were the Senior Writers for the first edition, published by It's About Time, Inc., of Armonk, New York. The assessment specialist was Don Hastings of the Stratford, Connecticut, public school system. Senior Writers for the second edition are William Berlinghoff and Karen Santoro (Central Connecticut State University), working in conjunction with Barbara Zahm (the project director) and Monica Rodriguez of It's About Time.

1. The work on which this chapter is based was funded in part by the National Science Foundation (NSF) through Grant No. ESI-9255251, awarded to the Connecticut Business and Industry Association. All opinions, findings, conclusions, and recommendations expressed here are those of the author and do not necessarily reflect the views of the NSF.

- Senior Writers—four college or university professors with extensive mathematics teaching and writing experience, who were charged with developing and writing drafts of all the books

- Contributing Writers—seventeen high school and community college mathematics teachers and two science specialists, who would review the chapter drafts and write additional problems, projects, and ancillary materials

- Advisory Council—twenty respected professionals from the fields of mathematics, science, education, and business—including several representatives from CBIA member companies—who would meet annually to review the overall direction of the project and to provide advice about various aspects of the emerging materials. A particular charge of this group was to comment on the extent to which these materials reflected realistically the skill and problem-solving expectations of the world of work.

- Pilot- and Field-Test Teachers and Students—more than 100 field-test teachers and approximately 2,500 students from seventeen different inner-city, urban, suburban, and rural high schools, who would use preliminary versions of these materials and provide feedback from their classroom experiences

- Evaluator—an independent consultant with background in the evaluation of educational projects, who would report to the Advisory Council and the principal investigators (PIs) about students' performance and the overall effectiveness of the project

The guiding vision of the project was that *all* students can learn mathematics, be critical thinkers, and be problem solvers. The mission, then, was to develop each student's conceptual understanding so that this vision would become a reality for every one of them.

Design Principles

MATH *Connections* was designed to implement *Curriculum and Evalation Standards for School Mathematics* (National Council of Teachers of Mathematics [NCTM] 1989) with the following principles in mind:

- The curriculum is intended for *all* students, regardless of their future education or career plans. It is relevant to students as future citizens, parents, voters, consumers, researchers, employers, and employees. It serves the needs and interests of those who will go on to further mathematics-intensive studies and science-related careers, as well as those who choose to pursue other fields of study or enter the workforce directly after high school. This is reflected, in part, by a two-tiered approach to the content—an "inner core" of material considered essential for students of all ability levels and an "outer core" encompassing more challenging aspects of the topics covered.

- It is flexible, affording students with different learning styles equal opportunity to master the ideas and skills presented. This means providing both interactive group work and individual learning experiences, encouraging frequent student–teacher and student–student interaction in finding answers and in discussing mathematics and students' relevant experiences, opinions, and judgments.

- It employs the main technological tools of the world of work. Students must learn to be comfortable with graphing calculators and computers in order to cope with and profit from the opportunities of our increasingly technological world.

- It is reality-based. All the mathematical ideas in the curriculum are drawn from and connected to real-world situations, with care taken to avoid any oversimplifications that destroy the realism of the results. Students can immediately see how the mathematics they are learning relates to their own lives and the world around them.

- It focuses on mathematical processes—reasoning, pattern seeking, problem solving, questioning, and communicating with precision—because those features of a mathematical education are as important to lawyers, doctors, business leaders, teachers, politicians, social workers, military officers, entrepreneurs, artists, and writers as they are to scientists or engineers.

- As its name suggests, **MATH** *Connections* explores connections of all kinds: between mathematics and the real world of people, business, and everyday life; between mathematics and the sciences; between mathematics and other disciplines, including history, geography, language, and art; and among different areas of mathematics.

Program and Course Design

MATH *Connections* is a unified approach that blends traditional mathematical topics around common thematic threads. The decisions about what topics to include and how to organize them were made during a weeklong meeting of the PIs and the Senior Writers in September 1992. At that meeting, we consulted the recommendations of the NCTM *Standards* (1989) and the addendum, *A Core Curriculum: Making Mathematics Count for Everyone* (Meiring et al. 1992). After deciding which topics we would include in the three years of our curriculum, we set about the task of ordering and connecting them.

Several early decisions, based on the principles above, guided that organization. The "inner core/outer core" structure allowed this curriculum to be taught in classes determined by skill level, as well as in heterogeneous classes. However, since this program was likely to be very different in style and approach from the previous mathematical experiences of many students, it was hoped that some students in inner-core classes might well want and need to be moved into an outer-core class. In part to allow such a move to happen in midyear and in part to provide other flexibility in implementation, the three-year curriculum was separated into six half-year books. Each book would assume only the inner-core material from previous books. A Teacher's Guide would accompany each of those six books.

Far more than simple answer books, these Teacher's Guides were to be detailed commentaries on the content and pedagogy of the text materials, particularly in those areas not usually included in a traditional high school curriculum. Their descriptions of the goals, options, and potential pitfalls of each section would allow teachers to see each concept, each section, and each chapter as the authors saw them when they were being written.

From the perspective of 1992, it seemed appropriate to take a two-pronged approach to technology by building in graphing calculators and computer spreadsheets as tools to be used routinely throughout the curriculum. To this end, appendixes on these two technologies would be included in each book, for ongoing reference purposes.

Another global decision was that the curriculum would be built around conceptual "threads" or strands rather than around problems. The content of each book has a unifying theme:

Year 1— Data, Numbers, and Patterns

Year 2— Shapes in Space

Year 3— Mathematical Modeling

Within those general themes, each chapter has its own unifying "story line." The chapters are large (there are only twenty-two in the entire program), each one presenting a major topic in a way that fits into the theme of its book.

The four chapters of the first book typify these organizational principles at work. The thematic goal, realized in chapter 4, is the calculator-assisted use of linear regression to solve some real-world prediction problems. The rest of the book is organized around reaching this goal: single-variable statistical ideas in chapter 1, first-degree equations in chapter 2, and graphs of linear equations in chapter 3. Practice in using a graphing calculator is built in along the way, so that by the time students reach chapter 4, they have all the tools needed to deal with the problems posed.

Many of the topics from algebra, statistics, combinatorics, and probability introduced in Year 1 are revisited in a broader, deeper, more sophisticated way in Year 3. Moreover, expectations of students' ability to justify, explain, and communicate their ideas move gradually from informal arguments and heuristic explanations in Year 1 to a more formal, logical argument structure by the end of Year 3. The goal of this reasoning thread is twofold. First, we want students at every level to acquire the habit of thought that all the mathematics they do must "make sense," to them and to others. Second, we want to lead students to a comfortable understanding of the process of deductive proof and its role as the backbone of formal mathematics. This is done by increasing in stages the sophistication of the explanations we present and expect, from the intuitive informality of Year 1, through the more carefully reasoned derivations and justifications of Year 2 and the first half of Year 3, to the principles of logic and their use in proving sequences of theorems in various axiomatic settings, including but not limited to Euclidean geometry, in the second half of Year 3. In our view, this gradual development appropriately reflects the increase in students' intellectual maturity as they progress from ninth to eleventh grade.

The issue of intellectual maturity also guided a major decision about our approach to geometry. Rather than mix the informal and the axiomatic approaches to geometric topics, we opted to present geometric ideas informally in Years 1 and 2, deferring any discussion of formal proofs until the last half of Year 3. At that time, axiomatic structure and formal proofs are introduced, not as something peculiar to geometry, but as an underlying logical structure of mathematics as a whole. Then, in the last chapter of Year 3, the structure of Euclid's geometry is explored from both historical and logical points of view.

Designing the Chapters and the Sections

Chapters are the major topical units of the **MATH** *Connections* program. Each chapter is divided into sections, and each section presents a subtopic that fits into the "story line" (i.e., the development of the mathematical topic) of the chapter. The size of each section is determined by how long it takes to develop its topic to some sort of natural conceptual conclusion. Thus, sections tend to be fairly long, often containing far more material than can fit into a typical class period.

To ensure students' ongoing, meaningful involvement in the development of each section's ideas, the following devices are used throughout the text:

- *Do This Now* identifies questions that students should deal with before moving ahead. Often these questions are fairly natural and will be answered in the following few pages of the text, but they will be much more meaningful if students engage them on their own first. Occasionally the questions simply reinforce a process or skill.

- *Discuss This* identifies questions intended to provoke interchange among students, either in small groups or in a whole-class setting. They often are open-ended, and some of them can lead to quite sophisticated ideas if pursued in depth.

- *Write About This* identifies writing exercises that often require students to gather additional information on their own or to reflect on a topic. The length of the writing expected varies from a few sentences to several paragraphs.

At least one of these instructional devices appears on virtually every page of every section. Many of them are best handled in a small-group setting. There is also a Problem Set at the end of each section, providing for a range of activities from routine practice to in-depth exploration and extension.

We knew from the start that the students would actually have to read these books! For some of them, and indeed for some of their teachers, reading a mathematics text would be an unfamiliar and perhaps uncomfortable experience. To ease some of that discomfort and to assist students in thinking through and communicating their mathematical ideas, three margin-note devices were designed:

- *About Words* explains how mathematical terms are related to words commonly used in everyday speech or writing.

- *About Symbols* points out how specific symbolic conventions are reasonable shorthand ways of communicating the ideas they represent.

- *Thinking Tips* highlight the use of specific problem-solving techniques, acting as reminders of useful "rules of thumb" for thinking about mathematical questions.

In addition, the reading level was restricted to approximately a year below the grade level for which the materials were intended.

Development Process

The actual writing of the course materials began in the fall of 1992. Most of the overall design described above was in place by then, although some of the details were shaped and refined during the following year. Each chapter was initially drafted by one of the Senior Writers. During the first six months or so, the Senior Writers met with the PIs weekly in Connecticut, usually at Wesleyan University in Middletown, to discuss the material being prepared. Those meetings became biweekly and then monthly in subsequent years. In each of the first three summers, the Senior Writers spent four weeks in residence at Wesleyan University, where they met daily with the team of Contributing Writers and the PIs. During these summer sessions, the draft chapters were refined, expanded, and prepared for pilot testing in the field-test schools. At the same time, our assessment specialist, aided by the Contributing Writers, was preparing assessments to be used by the field-test teachers.

Pilot testing began in the fall of 1993 in five high schools. The agreement with the administrations of each of the participating schools was that two teachers would each have two sections of **MATH** *Connections*, one inner-core section and one outer-core section. These two teachers would be given a common planning period during the school day so that they could discuss their implementation of the materials. The summer before teaching the program, the teachers would attend a two-week training session, held at Wesleyan University. In addition, there would be several Saturday follow-up sessions for the teachers during the school year. This same model was to be replicated in 1994 and 1995, with at least five new schools joining the field testing in each of those years, as the schools already in the program moved to the second- and third-year materials. Thus, by the end of the field testing, at least fifteen schools would have tried all three years of the **MATH** *Connections* materials at least once. (In fact, seventeen schools were actually involved in the pilot testing.)

Texas Instruments supplied classroom sets of TI-82 graphing calculators for use in the field-test classes, so this calculator model was used as the prototype for calculator material in the text, with some notes in the Teacher Editions about modifying the instructions for other calculators. An appendix describing how to use this calculator "from scratch" was written for inclusion in all the books. It was eventually extended to include the TI-83 when that model became available.

Throughout the preparation and testing of the text materials, particular attention was paid to questions of gender equity and ethnic equity. Consultants outside of the writing team reviewed the materials from these perspectives and found them to be appropriately sensitive to both areas of concern. Particularly gratifying to the writers in this regard were the data from the early rounds of field testing, which indicated that there were no significant gender-related differences in either performance or attitude.

During the years 1995–1998, an external team conducted an extensive evaluation. They examined students' test scores, observed program classrooms, and interviewed participating students and teachers. Overall, their findings confirmed that the program was performing as intended. For a detailed description of those findings, see Cichon and Ellis (2003).

As might be expected, not everything proceeded as planned. Some early, unforeseen changes in the initial Senior Writers team set the writing schedule back almost immediately. This caused considerable inconvenience from time to time, but did not affect the final product in any significant way. Three other kinds of difficulties were more troublesome:

- Some of the field-test schools did not adhere to the staff assignment model they agreed to. In some instances, the classes were heterogeneously grouped; in others, they were all inner-core. Some teachers did not have common planning periods, as promised. In all but one instance, these situations were accommodated without losing a test site, but the irregularities affected the quality and uniformity of the data collected.

- At a few field-test schools, small groups of parents who had been contacted by opponents to *Standards*-based curricula (generally under the self-proclaimed title "Mathematically Correct") objected to having their children included in the **MATH** *Connections* classes. This required the project PIs to meet with school boards and parents' groups, and in some instances resulted in schools setting up parallel courses, so that students or their parents could opt for a traditional curriculum.

- In an apparent misunderstanding of his role, the original external evaluator gathered data and prepared reports that were more aligned with his own area of academic research than with the goals of the program. After two years, he left the project on short notice. As a result, important evaluation data from the early stages of pilot testing was either missing or flawed.

A disappointment in the development process was the planned interaction with the business and industry members of CBIA. Despite their continued willingness to cooperate, we seemed to be unable to settle on a dialogue format that would generate for the writing team the kind of information needed to create problem scenarios and significant, well-motivated contexts for mathematical exploration. Although this was something of a missed opportunity, it did not affect the integrity of the program materials. The CBIA advisors who commented on the real-world settings created by the writing team did so quite favorably.

Another unforeseen complication was in the area of technology. During the first field-test year, many of the teachers reported that they were unable to get access to computers in their schools when they needed them. In response to this, the writers de-emphasized spreadsheet explorations and other computer-based material, focusing instead on the graphing calculator as the only required technology. Spreadsheet work and investigations with Geometer's Sketchpad or similar software were relegated to optional status, described in the Teacher Editions but not included in the student texts.

Feedback from the field-test sites did not require any other major changes in the content or structure of the program as originally planned. Of course, many minor adjustments were made in response to comments from the teachers and students. In addition, in preparation for the program's publication by It's About Time, Inc., of Armonk, N.Y., the writing process was extended in three ways:

- Each chapter was thoroughly revised and edited by a Senior Writer other than the original author (but, of course, in consultation with the original author) to give more uniformity and continuity to the final version.

- Some of the Contributing Writers and field-test teachers wrote supplementary materials to provide skill reinforcement in ways that are compatible with the **MATH** *Connections* text materials.

- Two of the Senior Writers wrote a set of forty-two Extensions, one or two for each chapter, that give able, motivated students opportunities to explore some concepts more deeply. These Extensions expanded the program's already rich resources for students at the upper end of the ability spectrum, affording teachers additional flexibility in providing challenging material that such students can work on, either independently or in small groups.

Lessons Learned and Implications for the Future

Reflecting on possible changes to our design principles brings to mind two things:

- The advances in technology during the dozen or so years since the inception of **MATH** *Connections* have brought with them opportunities for a wide range of educational activities that were unavailable in the early 1990s. This has been accompanied by greater access to that technology in many schools. These advances invite increased use of technology in the materials. Nevertheless, there is a paradoxical trap here for designers of secondary school curricula. Many postsecondary institutions, including some of our most prestigious universities, do not allow or encourage students' use even of calculators in their early mathematics courses. Thus, as we prepare all students to be more skilled users of technology, we must find a way to insulate those who aspire to higher education from being penalized for their unfamiliarity with a technology-free mathematical environment.

- It might be appropriate to extend **MATH** *Connections* to a four-year program. As mentioned in the first section of this chapter, **MATH** *Connections* was designed to be a common, three-year core on which further study of mathematics might be based. Students who wanted a fourth year of mathematics would be at the point of choosing among a variety of alternative courses on the basis of their interests and career aspirations. A fourth year was envisioned in the original grant proposal as one in which several options would be available to students from existing or future courses in their high schools. Nevertheless, keeping in mind that this curriculum is intended to be suitable for all students, a fourth year of **MATH** *Connections*, written in the same spirit as the first three, might provide a useful alternative for students, most of whom will not go on to study calculus at the postsecondary level. The threshold challenge, of course, is to choose content that is truly relevant to the widely varied audience it is likely to serve.

Several refinements to the development process would be in order. High on the list is more careful planning of the logistics of handling mathematical manuscripts from different authors using different text preparation programs. This extends to the obvious, essential step of transmitting material to the publisher in a way that minimizes typesetting difficulties and related production issues.

Of particular interest to this project, given its genesis through the CBIA, is the way in which we might interact with willing collaborators from business and industry. When we asked business or manufacturing people to furnish ideas for real applications of mathematics in their industry, most of them were unable to propose any worthwhile suggestions that fit into our curriculum. That should not have been a surprise. We should not have expected them to see their use of mathematics through our eyes. They didn't know our curriculum or students well enough to do that. Nor should they have; it wasn't their job. It might have been far more productive if the writers were given guided access to a business or industrial site for a day or a week (or longer), were allowed to observe what was going on around them, and then processed that information through the eyes of the curriculum to see where scenarios from that business made pedagogical sense.

Beyond the actual development of the text materials, the single biggest challenge is the professional development and support of teachers who use this program after the field-test stage is over. Our extensive Teacher Editions provide some help with unfamiliar content, but books alone cannot prepare teachers to implement this curriculum in the way it is intended. That preparation requires more personal attention, training, and ongoing support. Such support depends, at least in part, on the willing cooperation of schools, which often requires a favorable administrative climate. This may require curriculum developers to take a much more active role in convincing parents, prospective employers, universities, and other stakeholders that what we are doing is valuable and effective.

A necessary (but not sufficient) component of such efforts is the compilation of good, persuasive data showing that what we do works for what these broader constituencies want. Like it or not, we are functioning in an impatient society, a society that wants to see immediate results. Educational change needs time to become effective, but we don't have as much time as we would like to make our case. It is too easy for a few vocal critics to make people forget that the reason the NCTM *Standards* documents were written and the reason that our curricula were funded by the NSF is that the old ways didn't provide the education that they (not just we) wanted their children and our workforce to have. We need to make our case more persuasively and more persistently.

A final lesson learned is: Be prepared to cope with Murphy's Law.

REFERENCES

Cichon, Donald, and June G. Ellis. "The Effects of **MATH** *Connections* on Student Achievement, Confidence, and Perception." In *Standards-Based School Mathematics Curricula: What Are They? What Do Students Learn?* edited by Sharon L. Senk and Denisse R. Thompson, pp. 345–74. Mahwah, N.J.: Lawrence Erlbaum Associates, 2003.

Meiring, Steven P., Rheta N. Rubenstein, James E. Schultz, Jan de Lange, and Donald L. Chambers. *A Core Curriculum: Making Mathematics Count for Everyone.* Curriculum and Evaluation Standards for School Mathematics Addenda Series, Grades 9–12. Reston, Va.: National Council of Teachers of Mathematics, 1992.

National Council of Teachers of Mathematics (NCTM). *Curriculum and Evaluation Standards for School Mathematics.* Reston, Va.: NCTM, 1989.

13

The Case of
Mathematics: Modeling Our World

Sol Garfunkel

THIS chapter is an essay intended to be both anecdotal and truthful. Anecdotal, because formal descriptions exist elsewhere and have been repeated in countless reports and marketing documents. Truthful, because I believe that there is often pressure on curriculum development projects to present a neater picture of progress and practice than the actuality of the untidy real world. I firmly believe that the essential ingredients in a successful project are the project personnel, including the writing team, and a deep understanding of and commitment to the project's articles of faith. By articles of faith, I mean those principles that will be adhered to regardless of external factors, such as publisher or funder demands. On occasion, every curriculum development project must come back to an understanding of first principles, much like engaging in scientific research.

The Application and Reform in Secondary Education (ARISE) project began with the stated intent of creating a curriculum that taught mathematics through real and contemporary applications. Moreover, we took on the task of teaching students to be mathematical modelers, believing that the modeling process was every bit as important an object of study as any particular piece of secondary school mathematical content. To be clear, we believed (and still believe) that if we could not find, for a particular mathematical topic, a real problem to be modeled, then that topic would not be included in our curriculum. We also insisted that each chapter of our texts be organized around an area of application and contain the mathematics necessary to analyze and model problems in that application area. We understood from the beginning that this would create a nonstandard curriculum that might or might not be well received by the secondary school community. We believed that our job was to create a paradigm—a proof of concept if you

The project leadership team consisted of Henry Pollak, Teacher's College, Columbia University; Landy Godbold, Westminster Schools, Georgia; Kay Merseth, Harvard University; and Sol Garfunkel, COMAP. Jan de Lange and other members of the Freudenthal Institute worked with us on the design of assessments. Beatriz D'Ambrosio, Indiana University Purdue University Indianapolis, helped us with professional development for the broader field test.

will—that such a curriculum (*Standards*-based, as we read the *Curriculum and Evaluation Standards for School Mathematics,* published in 1989 by the National Council of Teachers of Mathematics [NCTM]) was possible. This also meant that we assumed maximum availability of technology for that time, that is, every student was expected to have a graphing calculator at home and access in every classroom to at least one computer with graphing utilities, spreadsheets, and, at the latter stages of the project, Internet access.

Instead of "strands" as they are usually defined, we chose to organize the curriculum around modeling themes such as Risk, Fairness, and Optimization. We made an explicit decision at the first authors' meeting not to create a grid with boxes for mathematical and application topics. Instead, within the themes we chose areas and problems that we believed would carry a good deal of the secondary school curriculum. Careful outlines were drawn and redrawn as the actual creation of the units took shape. For each course, we chose major mathematical themes to emphasize. For example, it was decided that one of the major mathematical themes of Course 1 was to be Linearity, so that each of the units in the course had to carry material leading to a deepening understanding not only of linear functions and equations, but also of the underlying concept of linearity.

As the text for each course neared completion, we held "guilt" sessions— meetings of the project staff and author team in which we asked ourselves how guilty we felt about not having included the standard mathematics topics for that particular grade—and whether we should find a way to include them. In the early years, when we worked on Course *n* and had these discussions, the answer was almost always to include the material in Course *n* + 1 and be thankful that we had four courses to play with. I should embellish on my use of the word *play*, because I believe that it is apt. If one takes, as we did, the position that we were creating a curriculum built to teach both mathematics and modeling, organized by theme and application, then there really was an enormous number of degrees of freedom with which to play. The constraints of which topic needed to be taught before which other topics was really more of an implementation problem than one of design—one which we expected the field test to help resolve.

The project leadership team worked closely with some thirty teachers, the overwhelming majority of whom were high school mathematics faculty. The search and application processes for these teachers were both thorough and extensive. I believe that we assembled one of the finest author teams possible to do this work. And one of the most important outcomes of this project is the success that these authors have enjoyed since. Many have become leaders in the mathematics education community, and I believe that their work on the ARISE program helped them realize their potential and offered them a missing point of entry.

ARISE was written in what we called "camp." Camp consisted of six-week (three weeks on, two weeks off, three weeks on) summer sessions for four years. Each unit was designed and written by a three-person author team. Everyone read the drafts, and we spent our nights in meetings discussing what we had read and what changes needed to be made. It was a remarkably democratic, though anarchistic, environment. Of course, some were more equal than others. Henry Pollak was the arbiter of both mathematical correctness (in the nonpolitical sense) and the honesty of the modeling examples. Landy Godbold was probably the only member of the team to understand everything in the curriculum, since he worked to ensure we would finish the project in finite time. Kay Merseth—and later, Beatriz D'Ambrosio—tried to keep us from overwhelming student and teacher alike. And I brought the Cokes and paid for the pizza. But nevertheless, we all had a say in what went into every unit, and we all knew, or at least had access to, each draft of each piece as it was being produced. In fact, we rented and broke two very large Xerox machines in the process.

The last week of camp each year was devoted to the professional development of the field-test teachers. These teachers joined us for the week, living in the dorms, sharing food and lack of sleep. We held sessions each day to make them aware of the course materials developed. Although there was follow-up work with the teachers during the school year, it was still true that they had to go through a year's worth of material in one week. And to be honest, in most years we weren't finished at the end of the camp: we were still polishing up the units before, during, and after the field-test teachers saw them. The good news about this process was that frequently the units were revised before they were done—taking into account the field-test teacher concerns—even in first draft form. The good news/bad news of the process is that after a summer or two, the field-test teachers became part of the family, part of the development team. What was the bad news here? I think it was twofold. First, the field-test teachers were largely self-selected. Not that we didn't have an application process, but these were teachers and schools who committed themselves to a three-year field test with materials yet to be created. They were perforce rather special people. In general, although their schools and students represented a broad range, the teachers were simply not your typical high school mathematics faculty. And as I said, they became family.

As a consequence, we got wonderful feedback on how to improve the curriculum we were making, but not necessarily on how that curriculum would be received and perceived by a less committed teacher or school system. Nevertheless, I would not have changed this process for the world. Also, many of our field testers have gone on to do significant and nationally recognized work in mathematics education. And whether their schools are using ARISE materials or not, their teaching has been changed forever. This last point cannot be emphasized enough. We are living through a time where it is becoming increasingly popular to say that we should take a break from curriculum development in order to research first how effective the past wave of development has been on students' performance.

But curriculum development is staff development, has always been staff development, and always will be staff development. Although curricula may come and go, the teachers remain. And if we can affect the teaching staff through exposure to new materials and approaches, those changes will have positive effects on students' performance far into the future. Although I cannot say that all the students who have studied Mathematics: Modeling Our World know and understand the modeling process, I can say that the faculty we have trained do, and that their modeling skills will be passed on regardless of what curriculum they teach.

Three members of the Freudenthal Institute came to camp each summer and worked directly with the author team and project staff in the development of the assessments. The assessments were generated first by members of the Dutch team and then discussed as part of the general discussion—unit by unit. I should also point out that the group from the Freudenthal Institute were themselves curriculum developers and as such quickly became part of the team—not simply restricted to the assessment program.

The last point that I would like to make is one about timing and schedule. No writer ever saw a draft he or she didn't want to revise, revise again, and then further revise. It is hard to let go, hard not to believe that with one more pass you can really get it right. But these projects had two forms of "drop-dead dates." The first was driven by the field test. We knew that in August—each year for four years—we had to deliver student-ready materials. Further, chapter 2 had to be ready in September, and so on.... This put enormous pressure on the writers and on the production staff. Four years' worth of field-testable materials is an enormous undertaking. And field test or no, these were real students who needed materials with accurate drawings in a format that promoted learning instead of getting in the way of learning. And we were typical authors. When we were done with the "intellectual" work of creating the materials, we expected them produced the next day. But deciding formats and elements for consistency (do we box definitions, and so on) is the work of professionals—demanding and creative—something to be valued and rewarded. Without the dedication of the Consortium for Mathematics and Its Applications (COMAP) production staff, we would never have been able to meet project deadlines.

The second deadline was actual publication dates and backing up to when final manuscripts had to be delivered—in our instance, directly to the printer, because we designed and produced the texts in-house. As in the instance of most of the projects in this volume, we produced our final texts one year at a time. Each year we bucked up against the deadline—cutting corners, scrambling to get the teachers' materials and supplements done in the same calendar year, and trying to keep the rising tensions and stress at an acceptable level. During these crunch times, compromises were made and forgotten. Once the texts were published, we went from being authors and project staff to becoming marketers. When you are selling, you don't focus on the blemishes and they are soon lost until the next edition.

When one does mathematical research, there is frequently an anticlimactic feeling when a new theorem has been discovered or proved. You wonder why it took you so long, why you didn't see such an "obvious" proof. Curriculum development is precisely the opposite. When you are creating the curriculum, you know that you don't know what you're doing. You're sure that every chapter should be rewritten. But when it's all done, when you see the bound book in two- or four-color, with the photographs and the graphics, it looks pretty good. And you really want every teacher to read it and every student to study from it. I am certainly no different in that respect from any other curriculum developer.

REFERENCE

National Council of Teachers of Mathematics (NCTM). *Curriculum and Evaluation Standards for School Mathematics.* Reston, Va.: NCTM, 1989.

14

The Case of the

Systemic Initiative for Montana Mathematics and Science (SIMMS)

Johnny W. Lott
James Hirstein
Gary Bauer

I N 1991, the Systemic Initiative for Montana Mathematics and Science (SIMMS) project was funded by the National Science Foundation to the Montana Council of Teachers of Mathematics.[1] Unlike other projects described in this volume, SIMMS was a statewide systemic initiative focusing on improving mathematics and science education through policy, curriculum development, and professional development. It had the following objectives (Lott and Burke 1993, p. 1):

1. To promote integration in science and mathematics education
2. To redesign the 9–12 mathematics curriculum for all students using an integrated, interdisciplinary approach
3. To develop and publish curriculum and assessment materials for grades 9–16
4. To develop a shared vision of K–12 science education
5. To incorporate the use of technology in all facets and at all levels of mathematics education
6. To increase the participation of females and Native Americans in mathematics and science

The SIMMS Integrated Mathematics: A Modeling Approach Using Technology curriculum was developed by the Montana Council of Teachers of Mathematics and is published by Kendall Hunt Publishing Company. The SIMMS project codirectors were Johnny W. Lott and Maurice Burke; the Materials Development Committee was chaired by Dean Preble and Terry Souhrada. Collaborators included writers from across the United States. Development of the third edition was codirected by Terry Souhrada and Peter Fong.

1. The work on which this chapter is based was funded in part by the National Science Foundation (NSF) under Cooperative Agreement No. OSR 9150055 to the Montana Council of Teachers of Mathematics. Any opinions, findings, conclusions, or recommendations expressed here are those of the authors and do not necessarily reflect the views of the NSF.

7. To establish new certification and recertification standards for teachers
8. To redesign teacher preparation programs using an integrated, interdisciplinary approach
9. To develop an in-service program on integrated mathematics to prepare teachers of grades 9–16
10. To develop the support structure for legislative action, public information, and general education of the populace necessary for effective implementation of new programs

For this article, we concentrate only on background for objectives 2, 3, and 5.

The SIMMS Curriculum Development Philosophy

As an introduction to the design of the curriculum, SIMMS codirectors led the search for a philosophy that could guide the curriculum. Considered were both instrumentalist and constructivist views as described in Davis, Maher, and Noddings (1990). The philosophy subscribed to by the project was consistent with those suggested by Resnick (1987), Skemp (1987), and others. The constructivist paradigm is consistent with the *Curriculum and Evaluation Standards for School Mathematics* as seen in the following (NCTM 1989, p. 10):

> This constructive, active view of the learning process must be reflected in the way much of mathematics is taught. Thus, instruction should vary and include opportunities for appropriate project work; group and individual assignments; discussion between teacher and students and among students; practice on mathematical methods; exposition by the teacher.

To adopt the constructivist view and to develop the curriculum, the project used the definition of integrated mathematics constructed by the Integrated Mathematics project (Beal et al. 1990, p. 13):

> An integrated mathematics program for all students is a holistic mathematical curriculum which:
> - consists of topics chosen from a wide variety of mathematical fields and blends those topics to emphasize the connections and unity among those fields;
> - emphasizes the relationships among topics within mathematics as well as between mathematics and other disciplines;
> - each year, includes those topics at levels appropriate to students' abilities;
> - is problem-centered and application-based;
> - emphasizes problem solving and mathematical reasoning;
> - provides multiple contexts for students to learn mathematical concepts;
> - provides continual reinforcement of concepts through successively expanding treatments of those concepts;
> - makes use of appropriate technology.

With the philosophy identified and the definition adopted, the SIMMS curricular goal was to develop a grades 9–12, integrated program rooted in applications and accessible to all students. The project was designed to promote increased attention on females, minorities, and disabled students. To this end the project (*a*) used multiple learning styles and delivery modes to address the needs of students of both genders, all races, and those with special disabilities; and (*b*) provided material and instruction to strengthen sensitivity to the value of multiple perspectives and the negative effects of bias and stereotyping.

The SIMMS project developed a framework of learning with a core of understandings considered vital to students' mathematical activities. The elements at the core are considered to be in continual evolution and include the following mathematical literacies: numeric, operational, functional, graphical, spatial, statistical, modeling, measuring, and computer/algorithmic. Not only were there core mathematical understandings, but core metamathematical attitudes were also identified (cf. Lott and Burke 1993, p. 9).

With the defined literacies and metamathematical attitudes outlined, the curriculum framework focused on four dimensions of mathematical activity viewed as critical components of the learning process. Those are mathematizing (defined by de Lange 1989, p. 100), problem solving, modeling (described by D'Ambrosio 1989), and integrating (defined in the work of Polya 1973, Skemp 1987, and van Hiele 1994).

Curriculum Levels

The SIMMS curriculum for grades 9–12 is divided into six levels, each consisting of one year of work. Within the levels, the curriculum is divided into modules of approximately two to three weeks in length. The modules were developed on the basis of a conical spiral of contexts with the mathematics concepts expanding each time that they are approached. It was proposed that every student would take the first two levels of the curriculum as seen in figure 14.1. After that, students have two options, Levels 3 and 4. The mathematical outcomes of Level 3 contain roughly half of Level 2 outcomes and approximately half of the outcomes of Level 4. Similar outcomes are seen in Level 5 coming from Levels 4 and 6.

Contexts

The SIMMS curricular framework was broadly divided into four types of contexts: mathematics and human systems, mathematics and environmental systems, mathematics and physical systems, and mathematical systems. The project used mathematical systems as the greatest single context area; all four contexts are seen at each grade level. In addition, the SIMMS project aligned its curriculum with the scope-and-sequence plan then being developed by the National Science Teachers Association (NSTA 1992). Scientific contexts were introduced when developmentally appropriate according to NSTA.

Technology

The SIMMS project was based on the belief that calculators and computers can be powerful assistants in the generation, organization, and analysis of many forms of data and many kinds of mathematical objects. In fact, the curriculum was built to incorporate graphing calculators, including the TI-92, with Cabri geom-

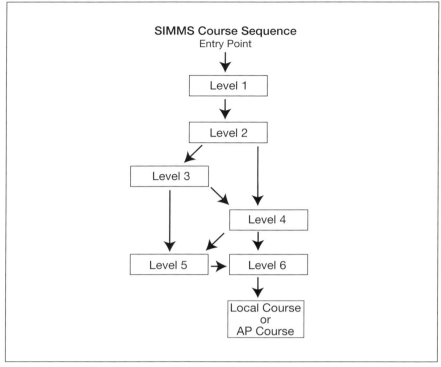

Fig. 14.1. SIMMS curriculum levels

etry and a computer algebra system, some statistical programs, a word process-ing program, and originally a potential computer-assisted design program. The classrooms designed to use the SIMMS project material were built on the notion of at least one computer for every four students. The SIMMS project, with the aid of the Montana state legislature, invested more than $1 million in technology for classrooms in the first two years of development. More technology was placed in the classrooms later.

Instructional Formats

The SIMMS materials were written to use a variety of instructional for-mats. All assumed heterogeneous classes. Included were formats of individual and cooperative-group work, whole-class discussions, and individual and group projects. Teaching materials allowed for individualization to specific needs and circumstances with the teacher expected to be a facilitator, motivator, and ques-tioner. Teachers play a pivotal role in shaping activities and harnessing curricular resources for both individual and group needs. The teachers were so important that teachers were primary writers of the materials. University mathematicians, mathematics educators, scientists, and specialist consultants from many different groups provided input for that writing.

Assessment

Assessment in the SIMMS project involved the coordination of, evaluation of, and assessment of all aspects of the project. This started with the initial curriculum and materials development through final implementation and testing in very large cities as required by the National Science Foundation.

The development of assessment materials for the SIMMS project was guided by the following principles (Lott and Burke 1993, p. 19):

- The restructuring of mathematics curriculum and instruction demands new assessment methods.
- Assessment should be an integral part of instruction.
- Assessment activities should be a learning experience for the students.
- Assessment should provide relevant information whose interpretation does not require an inordinate amount of time and effort.
- The type of assessment used should be aligned with the student outcomes expected.
- Assessments should document what students can do, not what they can't do.
- Assessment must document progress and processes as well as products.

Development Process

Because the SIMMS project was a state systemic initiative funded by the National Science Foundation, all the work of the project (and not just the curriculum) was developed around a plan for the entire state. The work of the project was directed by committees—Curriculum Development, Assessment, Professional Development, Government and Public Relations, a Steering Committee, and a National Advisory Committee—and a science director. Each committee played a vital role in the direction of the SIMMS project.

For the actual development of the materials, the Curriculum Development Committee with its chairs worked to outline the contexts and the mathematical concepts that would guide the writing. The original work started with a "white sheet" policy. That is, no absolutes were placed on the material to be written concerning what has been considered to be traditional material at the high school level. The major driving force at the onset of the project was "What mathematics should a mathematically literate adult know on graduation from high school?" Another major goal was to keep all mathematics options open to students for as long as possible in their high school careers. By developing the levels discussed above, the project tried to allow access to mathematics for all levels of students from the less-inclined ones to those who might be future mathematicians and scientists.

Writing was done primarily by teachers, with much of the writing done in the summers and rewriting and revising done during the academic years the project was in place. Approximately eighty-five writers contributed to the curriculum development. As modules were written, they were read and reviewed by teachers in the professional development component of the project, were tested during the academic years in schools with trained teachers and using equipment purchased for the schools through the project, and were revised on the basis of the assessment of students and on the recommendations of the teachers piloting the project materials.

The writing, testing, revising, and retesting of the material continued for eight years. Originally the material was tested in Montana with few pilot classes outside the state. In the last two years of the extended project, materials were tested in Texas and in Ohio in large cities with diverse populations.

As the project developed, criteria were developed and used to evaluate the following (Lott and Burke 1993, pp. 25–26):

1. Implementation
   - the readability of written materials
   - the appropriateness of instructional contexts
   - the potential of materials, technology, assessment items, and instructional methods to be implemented by teachers and students in the intended manner

2. Mathematical content of materials
   - the appropriateness of level of assumed prerequisite knowledge
   - students' specific cognitive and attitude outcomes
   - students' abilities to transfer learning to other contexts and subjects
   - the development of students' higher-order thinking abilities
   - the development of students' ability to communicate mathematics

3. Impact on attitudes
   - students' and teachers' reactions to materials, assessment, instructional methods, and technology
   - students' and teachers' attitudes and beliefs about mathematics, teaching, and learning

4. Alignment with SIMMS project goals (determined through review by outside experts)
   - the integration of mathematical topics
   - the use of human and natural sciences contexts
   - appropriate coverage of essential mathematical topics

5. Development of assessment items
   - the use of multiple approaches
   - the alignment of items with specific content
   - the assessment of content, attitudes, thinking skills, communication
   - the appropriateness of evaluation mechanism.

A summary of results is reported by Senk and Thompson (2003). A long-term study of students who studied the SIMMS curricular materials, traditional materials, and a mix of the two different types of materials is reported in Souhrada (2001).

Conclusion

Changes were implemented at the very beginning of the project on the basis of a review by the National Advisory Committee. That review changed the nature of the levels and led to one for heterogeneous classes of students and the six levels of curriculum developed.

Current revisions of the material include more review material and more exercises for practice, and are condensed into four levels. The developers of the project would not change the use of technology, the deletion of some traditional mathematical topics, and a continuation of an integration of all mathematics. We believe that the evidence strongly favors the work that was done. With teachers who have been professionally trained and have used the technology and teaching methods recommended, all levels of students achieved as well on traditional mathematics tasks—and greater in the area of problem solving—using more varied approaches to problems than students in traditional curricula did.

REFERENCES

Beal, Jack, Dan Dolan, Johnny W. Lott, and John P. Smith. *Integrated Mathematics: Definitions, Issues and Implications.* Helena, Montana: Montana Council of Teachers of Mathematics, 1990.

D'Ambrosio, Ubiratan. "Historical and Epistemological Bases for Modelling and Implications for the Curriculum." In *Modelling, Applications, and Applied Problem Solving: Teaching Mathematics in a Real Context,* edited by Werner Blum, Mogens Niss, and Ian Huntley, pp. 22–27. Chichester, England: Ellis Horwood, 1989.

Davis, Robert B., Carolyn A. Maher, and Nel Noddings, eds. *Constructivist Views on the Teaching and Learning of Mathematics. Journal for Research in Mathematics Education* Monograph #4. Reston, Va.: National Council of Teachers of Mathematics, 1990.

de Lange, Jan. "The Teaching, Learning, and Testing of Mathematics for the Life and Social Sciences." In *Applications and Modelling in Learning and Teaching Mathematics,* edited by Werner Blum, J. S. Berry, R. Biehler, Ian D. Huntley, G. Kaiser-Messmer, and L. Profke, pp. 98–105. Chichester, England: Ellis Horwood, 1989.

Lott, Johnny W., and Maurice Burke, eds. *The Systemic Initiative for Montana Mathematics and Science (SIMMS) Project: Monograph 1: Philosophies.* Missoula, Montana: SIMMS Project, 1993.

National Council of Teachers of Mathematics (NCTM). *Curriculum and Evaluation Standards for School Mathematics.* Reston, Va.: NCTM, 1989.

National Science Teachers Association. *Scope, Sequence and Coordination of Secondary School Science. Volume 1: The Content Core, Guide for Curriculum Designers.* Washington, D.C.: National Science Teachers Association, 1992.

Polya, George. *How to Solve It.* Princeton, N.J.: Princeton University Press, 1973.

Resnick, Lauren B. *Education and Learning to Think.* Washington, D.C.: National Academy Press, 1987.

Skemp, Richard R. *The Psychology of Learning Mathematics.* Mahwah, N.J.: Lawrence Erlbaum Associates, 1987.

Senk, Sharon L., and Denisse R. Thompson, eds. *Standards-Based School Mathematics Curricula: What Are They? What Do Students Learn?* Mahwah, N.J.: Lawrence Erlbaum Associates, 2003.

Souhrada, Terry A. "Secondary School Mathematics in Transition: A Comparative Study of Mathematics Curricula and Student Results." Ed.D. diss., University of Montana, Missoula, 2001.

van Hiele, Pierre M. "A Child's Thought and Geometry." In *English Translation of Selected Writings of Dina van Hiele-Geldof and Pierre M. van Hiele,* published as part of the research project, "An Investigation of the van Hiele Model of Thinking in Geometry among Adolescents," edited by Dorothy Geddes, David Fuys, and Rosamond Tischler, pp. 243–52. Research in Science Education (RISE) Program of the National Science Foundation, Grant No. SED 7920640. Arlington, Va.: National Science Foundation, 1994. (Original work published in 1959)

15

The Case of the

University of Chicago School Mathematics Project—Secondary Component

Zalman Usiskin

B EFORE the University of Chicago School Mathematics Project (UCSMP) began in 1983, this author had helped develop textbooks for each of the four high school years (Henderson, Usiskin, and Zaring 1971; Coxford and Usiskin 1971; Usiskin 1975, 1979). The algebra, geometry, and advanced algebra texts were characterized by innovative approaches (for the time) to the subjects, with strong emphases on transformations in geometry; applications in first-year algebra; and transformations, matrices, and groups in second-year algebra. The texts had also undergone significant testing. But except for the geometry and advanced algebra texts, the four books were not related to one another. The promise of six years of significant funding from the Amoco (now BP) Foundation made it possible not only to continue this work but also to think about a coordinated curriculum for all grades K–12. The size of the task led the UCSMP organizers to split the curriculum development task into two roughly equal components: an elementary (grades K–6) component, discussed in chapter 2 of this volume, and the secondary (grades 7–12) component, discussed here.

This chapter identifies some of the design principles underlying the work of the secondary component from 1983 until the present. This summary concentrates on principles of curriculum development. It is not an exhaustive list and purposely does not include principles of evaluation during that development or principles of marketing once the materials have become available. A history of the development and testing of each book in the curriculum is given in the Professional Sourcebook section of the Teacher's Edition for that book. A summary of the testing process can be found in Senk (2003).

Design Principles

Identification of the Task

The project was funded with the hope that the mathematics performance of students would improve. Projects in the 1960s and 1970s had often designed interesting materials that had minimal use. Change does not occur merely because we want it to occur; materials must incorporate enough of a different approach to enable desired changes to happen. The materials must be seen as implementable for use in classes, and when used, the materials must be perceived by users as effective in creating the desired changes.

- *To maximize interest in the curriculum that is developed, design the curriculum to confront issues that have been identified by others rather than issues the project itself has formulated.*

  Because of the large number of reports during the years 1975–1983 just before UCSMP began, we made a decision not to bring together people to create another report to guide us with regard to the content of a new curriculum (Usiskin 1985). Instead, we created a curriculum that would combine those recommendations of the reports that went along with updating the curriculum and upgrading students.

- *The beliefs underlying the curriculum need to be explicitly stated to guide authors and users.*

  Two fundamental beliefs underlay the development of the first edition's materials. First was the belief that students were not performing well enough. Second was the belief that the curriculum was out of date. We called these goals *upgrading students* and *updating the curriculum*. A third fundamental belief underlay the development of the second edition: curricular materials had to be broad enough to reach virtually all students (National Research Council 1989; National Council of Teachers of Mathematics [NCTM] 1989). These beliefs remain the foundation of the third edition.

- *If something is expected to be learned, it needs to be explicitly taught; if a topic is not in the curriculum, then it will not be taught.*

  Almost all curricula in the 1970s and early 1980s taught skills before applications (if any) of the skills. There was a naïve belief that a student who was skillful and knew mathematical properties of objects and operations would automatically transfer to being able to apply these properties and use these skills in problem situations. This belief was not supported by evidence (Begle 1973) and implied that if we want to improve students' ability to apply mathematics, we have to teach the applications.

The Development Process

We had used a three-year development cycle on three previous books that we had developed: a pilot year during which only authors and a handful of others use the materials; a formative evaluation year during which teachers using the materials are asked for their opinions on every lesson; and a summative evaluation year during which teachers are given the materials to use without much intervention from the project and the materials are compared with other materials.

- *A curriculum is only as good as the people who write it.*

  We had seen many ideas fail not because the idea was bad, but because the implementation of the idea was faulty. As a result, people would conclude—incorrectly, in our opinion—that the idea was not worthwhile from the start. It is not true that good ideas will overcome poor writing. We searched nationwide for people with both good ideas and good writing skills.

- *Talent for writing for students exists both in people teaching at the college level and those teaching at the school level.*

  We had worked with talented schoolteachers who knew more mathematics than most mathematicians as well as with talented mathematicians who understood school mathematics. College faculty tend to have more opportunity to cultivate particular curriculum research interests, whereas schoolteachers have more opportunity to try out new things in the classroom. The ability to write for students is not limited to one level or the other. So we strove for a combination of both college and school faculty on our writing teams.

- *An optimal way to develop good materials is for authors to teach the materials as they write.*

  This principle was applied in earlier works (Coxford and Usiskin 1971; Usiskin 1975, 1979). The next best way is for authors to teach the materials after they write them. Thus, first drafts of the materials were taught either by the authors themselves or by good friends of the authors so that they would get continual feedback for revisions.

- *Designing a curriculum for research purposes is different from designing a curriculum for universal use in schools.*

  Curriculum developers are constantly having to make choices of approaches to topics. Given a choice between a new approach that seems to work as well as an old approach, if one is doing a research study, one incorporates the *new* approach because of the desire to know more about the new approach; but in materials for use in schools, one should use the *old* approach because there is no sense in forcing change if there is no discernible difference between old and new. New approaches must be better than old approaches in order to be accepted by teachers. Why should a teacher learn a new approach if it doesn't do things better? For example, very early on we brought in people to design an integrated geometry and second-year algebra curriculum over two years. But the syllabus they created was essentially a half year of algebra followed by a full year of geometry finished off with another half year of algebra. Not wanting to fight the battle of implementation, we decided to make certain that all the algebra needed for the geometry was done in the first-year course. This left us with three courses (Algebra, Geometry, and Advanced Algebra) with traditional titles, even though they were quite integrated in content.

A Six-Year, Not a Four-Year, Curriculum

Although some improvement in students' performance might be achieved by better teaching of existing content, substantial improvement requires that a curriculum take advantage of what students know, avoid unnecessary review, and spend enough time on new content so that competence with that content is a reasonable expectation.

- *With a strong grade 7 course, algebra would be a reasonable course for the majority of students in grade 8.*

  In the 1980s, grade 8 mathematics textbooks were similar to grade 7 books, and grade 7 books were quite similar to grade 6 books (Flanders 1987). If a student did not need substantial review of the mathematics learned in prior years, algebra would be a natural eighth-grade course (Usiskin 1987). We further believed that if there were a strong K–6 program, then many students would be ready for algebra at grade 7, though they and their schools might not be willing to accept the amount of homework that successful completion of algebra requires.

- *Many students enter the typical one-year geometry course with little prior knowledge; this lack of knowledge is a key reason why they become unsuccessful at proof in geometry.*

  From earlier work (Usiskin 1982; Senk 1985), we knew that many students enter geometry with little prior knowledge and that the extent of prior knowledge was quite predictive of success in learning proof in geometry. Also, the previous geometry experience of those students tended to be detached from the other mathematics that students encounter. Consequently, we endeavored as much as we could to make geometry a strand throughout the curriculum and integrate it with the other mathematics students learn.

- *When we began our work, most students who entered college even with four years of high school mathematics were unprepared for calculus, a condition that persists today.*

  Before 1960, calculus was commonly a sophomore course in college even for mathematics majors. With the mathematics reforms of the 1960s, separate courses in analytic geometry and theory of equations, often covered before calculus, were eliminated from most college curricula, and as a result many students have only one year of mathematics between a second-year algebra course and calculus. Since that time, although we have evidence that many students can succeed in calculus on such a timetable, we also have ample evidence from the number and percent of students who require remedial mathematics in college that one year between a second-year algebra course and calculus is not sufficient to prepare most students for calculus (Waits and Demana 1988). With the need for all students to know more statistics and for college-bound students to be prepared to use some discrete mathematics, a decision was made to have two full courses in the UCSMP curriculum after second-year algebra and before calculus. Thus, students who went through the curriculum and took UCSMP Algebra in grade 8 would still not take calculus in grade 12.

Integrating Content

It may seem obvious that how one approaches the mathematics of the course can significantly affect how one thinks about that mathematics, but our earlier work had taught us that these effects are far greater than is usually recognized. In this section we concentrate on broad macrocurricular ideas, but the effects of curriculum are also significant for the ways in which individual concepts (slope, angles, multiplication) are treated.

- *Geometric transformations (rotations, reflections, translations, size changes) are required for a thorough study of geometry and enhance the study of algebra and functions.*

  Our earlier work with transformations in both geometry and algebra (Coxford and Usiskin 1971; Usiskin 1972, 1975) showed us the power that transformations have to change how students perceive geometric figures and relationships and how they approach functions and their graphs. Consequently, we integrated transformations into all courses.

- *Statistics naturally integrates with elementary algebra and functions.*

  Earlier work (Usiskin 1979) had convinced us that elementary algebra is strengthened by work with statistics. As we began examining how we would integrate statistics into later courses, we were struck by the many ways in which statistics and functions worked together. Thus was born a course ultimately called Functions, Statistics, and Trigonometry.

- *The understanding of mathematics involves at least five dimensions, of which four are important for student learning.*

  Understanding is multidimensional in the sense that one can have an understanding of a concept in any one or more of the dimensions identified below and not have an understanding of the same concept in the other dimensions. For example, one can have skill in and knowledge of the mathematical properties of an operation such as multiplication without a knowledge of the uses of multiplication, or its representations, or its history. We also mean that there is a range of understandings within each dimension from simple rote understandings to the invention of new understandings. We decided that each concept should be discussed from the perspectives of the first four dimensions of understanding, and some from the perspectives of all five dimensions.

  a. Skill-algorithm dimension—ranging from the carrying out of simple algorithms (mentally, with paper and pencil, or with technology) through the carrying out of algorithms with many decision-nodes to the invention of new algorithms

    b. Properties–mathematical underpinnings dimension—ranging from the recognition and application of individual properties to more complex justifications to the writing of original proofs

    c. Uses-applications dimension—ranging from the applying of operations of arithmetic in everyday problems to the use of more complex models to the invention of new applications for known mathematics or the invention of new mathematics for given applications

    d. Representations-metaphors dimension—ranging from concrete representations of abstractions through geometric representations (e.g., graphs and networks) of algebraic objects and logical relations and through algebraic representations (e.g., coordinates) of geometric objects to the invention of new representations (e.g., glyphs and box plots in the past twenty-five years) and metaphors

    e. History-culture dimension—ranging from being able to name dates and people to the knowledge of the evolution of mathematical ideas and the ability to re-create the mathematics of a given time or given culture (Usiskin 1991)

- *The teaching of applications of mathematics should be carefully sequenced over the entire curriculum.*

  The dimensions of understanding suggest one reason why students have so much trouble applying mathematics. Traditional curricula carefully sequence algorithms and carefully sequence mathematical properties, but no such sequence exists for applications. Moreover, students are typically shown generalizations that indicate the fundamental relationships between (pure) mathematics and its applications. For instance, typically it is only in the study of counting problems and probability that students see articulated the fundamental counting principle of addition that they used in first grade. We view principles of modeling such as these to be of equal importance as the assumed (theoretical) properties of mathematical concepts; without these principles there is no formal way to connect mathematics with its uses. From these basic principles, both simple and more complex applications of mathematics can be addressed (Usiskin 1989, 1991).

- *The latest in technology should be used, but only if the technology is available from more than one manufacturer.*

  We are cheating our students if we do not allow them the use of the latest technology. However, technology changes so quickly that a "cutting edge" technology at the beginning of the development of a curriculum will typically have been replaced by something better by the time the curriculum is ready for wide distribution. If the ideas behind the technology do not attract more than one manufacturer, then it is doubtful that the technology will be attractive enough to enough potential users to make it worthwhile, and if that one manufacturer decides to leave the market, then the curriculum will have no backup. In each edition of UCSMP, because the technology has become more powerful and available in a greater number of platforms, applying this principle has become more complex.

Text Features

In elementary school mathematics, virtually every skill and concept is discussed repeatedly in the class, and a child who does not learn an idea one day will have many more opportunities to pick it up. When students enter middle school and later in high school, classes may involve active participation, but homework becomes a significant variable in how much students learn, and students need to learn how to learn on their own.

- *People learn better if they are active learners, doing the exploring, drawing, solving, and discussing, than if they are passively looking at someone else doing the work.*

  Educators have realized the value of active learning since at least the time of Pestalozzi two centuries ago. But if a student does only worksheets and does not engage in exploration and discuss work with others, then the student can easily be practicing mistakes instead of good mathematics. In the second edition, we included activities in the book to be done in class as well as activities to be done at home. The third edition includes even more activities designed to place the learner inside the mathematics.

- *In order to learn to learn, students must learn to read mathematics.*

  Although there are classic examples of students ignoring words in problems, students typically read in order to solve a mathematics problem, and some students will read examples to mimic them. But in order to learn to learn mathematics, students must learn to read explanations as well as problems. Reading is also necessary to learn about the history of a concept, how the concept relates to other concepts, and how the concept can be applied.

- *Mastery learning works, but in practice it is almost impossible to implement.*

  Mastery learning is a strategy whereby students are taught some content, given a formative test over that content, given feedback and additional teaching on the concepts they answered incorrectly, tested again, given feedback, and tested, until they reach mastery of the content. The logic behind the mastery-learning strategy is that top students master content and then go on to the next material with that mastery, whereas slower students do not master the content, and when they go on to the next material, they are destined not to master it either. The research indicates that when slower students are allowed enough time to master content, they become almost like top students. The problem is that it can take up to five times as long for slower students to master content as faster students (see Bloom 1976, p. 188). As a consequence, if teachers wait until all students master topics, they will cover far less material in a course and actually have lower performance as a result. We found that a modified mastery-learning strategy, whereby students (1) tackle a chapter's worth of material at a time during which there has been continual review; (2) take a self-test to determine which objectives they have mastered and which they haven't, with *solutions* in the book providing the feedback; (3) follow the self-test with review exercises as correctives matched to the objectives; and (4) have continual review of earlier content for retention, is a powerful and implementable strategy.

- *If a topic is not tested, it will likely not be learned well.*

  We had seen materials in which applications were used as motivation, or theory used to connect ideas, or representations used for understanding, but these aspects were not tested. This principle implied (1) that student materials would have objectives, test for mastery of the objectives, and offer review exercises in all four dimensions of understanding and (2) that ancillary materials supplying additional practice, extra forms of tests, and extensions would be as broadly conceived as the text itself.

- *Projects can be a valuable way to capture the imagination and creativity of students.*

  In the first edition, by the time that students reached the fifth and sixth courses in the UCSMP secondary school curriculum, it was believed that they would be ready to tackle mathematical projects, that is, explorations into mathematics that might take a few days or longer to complete. Consequently, the end of each chapter of the first edition of these last two books (courses 5 and 6) included a description of five to ten possible projects that students might tackle. This idea was met with favor during the field tests, and with so much favor in the use of the hardcover book that in the second edition we included projects in all the books. Although we think that most teachers do not incorporate the projects into their assignments because of a lack of time, our experience is that those who do incorporate projects believe students find them to be exceedingly motivating and valuable and an outlet for their imagination and creativity.

The Current Scene

The appearance of the NCTM *Curriculum and Evaluation Standards* in 1989 spawned, through the NSF-supported projects of the 1990s, perhaps a greater variety of materials to be available than ever before in mathematics education in the United States. It is ironic that the No Child Left Behind legislation, which placed more authority in the hands of the federal government than ever before, has forced states to create their own curriculum standards and has resulted in a greater variety of mathematical expectations for students than ever before, and no curriculum can hope to meet them all. Furthermore, as expectations increase, the differences between the haves and the have-nots increase, and it becomes more difficult to create a curriculum that takes advantage of the advances we have made over two decades and still is able to be understood by those who have sat on the sidelines.

Many educators and the mass media act as if the standards movement in mathematics education in the United States has been a failure. This interpretation of the current scene ignores the remarkable increases in the performance of students in the United States in mathematics in the past twenty-five years—increases visible through longitudinal studies of the National Assessment of Educational Progress, through the examination of SAT and ACT scores through the years, and even through our showing relative to other countries on international tests. It ignores the fact that students are leaving high school having learned more mathematics than their counterparts of one and two generations ago. These mistaken interpretations of the current scene put pressure on us and others who are developing curricula to do away with some of the features of our curricula, such as the use of the latest in technology, that we think have fueled these increases in performance. In the development of the third edition of the UCSMP secondary curriculum, we cannot please everyone, but we also cannot with good conscience back away from the principles that we believe have worked so well.

REFERENCES

Begle, Edward G. "Some Lessons Learned by SMSG." *Mathematics Teacher* 66 (March 1973): 207–14.

Bloom, Benjamin S. *Human Characteristics and School Learning*. New York: McGraw-Hill Book Co., 1976.

Coxford, Arthur, and Zalman Usiskin. *Geometry—a Transformation Approach*. River Forest, Ill.: Laidlaw Brothers, 1971.

Flanders, James R. "How Much of the Content in Mathematics Textbooks Is New?" *Arithmetic Teacher* 35 (September 1987): 18–23.

Henderson, Kenneth B., Zalman Usiskin, and Wilson Zaring. *Precalculus Mathematics*. New York: McGraw-Hill Book Co., 1971.

National Council of Teachers of Mathematics (NCTM). *Curriculum and Evaluation Standards for School Mathematics*. Reston, Va.: NCTM, 1989.

National Research Council. *Everybody Counts: A Report to the Nation on the Future of Mathematics Education*. Washington, D.C.: National Academy Press, 1989.

Senk, Sharon L. "How Well Do Students Write Geometry Proofs?" *Mathematics Teacher* 78 (September 1985): 448–56.

———. "Effects of the UCSMP Secondary School Curriculum on Students' Achievement." In *Standards-Based School Mathematics Curricula: What Are They? What Do Students Learn?* edited by Sharon L. Senk and Denisse R. Thompson, pp. 425–56. Hillsdale, N.J.: Lawrence Erlbaum Associates, 2003.

Usiskin, Zalman. "The Effects of Teaching Euclidean Geometry via Transformations on Student Achievement and Attitudes in Tenth-Grade Geometry." *Journal for Research in Mathematics Education* 3 (November 1972): 249–59.

———. *Advanced Algebra with Transformations and Applications*. River Forest, Ill.: Laidlaw Brothers, 1975.

———. *Algebra through Applications*. Reston, Va.: National Council of Teachers of Mathematics, 1979.

———. "Van Hiele Levels and Achievement in Secondary School Geometry." Final report of the Cognitive Development and Achievement in Secondary School Geometry Project, June 1982. (ERIC Document Reproduction Service No. ED220288)

————. "We Need Another Revolution in Secondary School Mathematics." In *The Secondary School Mathematics Curriculum,* 1985 Yearbook of the National Council of Teachers of Mathematics (NCTM), edited by Christian R. Hirsch, pp. 1–21. Reston, Va.: NCTM, 1985.

————. "Why Elementary Algebra Can, Should, and Must Be an Eighth-Grade Course for Average Students." *Mathematics Teacher* 80 (September 1987): 428–38.

————. "The Sequencing of Applications and Modelling in the University of Chicago School Mathematics Project (UCSMP) 7–12 Curriculum." In *Applications and Modelling in Learning and Teaching Mathematics,* edited by Werner Blum, J. S. Berry, R. Biehler, Ian D. Huntley, G. Kaiser-Messmer, and L. Profke, pp. 176–81. London: Ellis Horwood, 1989.

————. "Building Mathematics Curricula with Applications and Modeling." In *Teaching of Mathematical Modelling and Applications,* edited by Mogens Niss, Werner Blum, and Ian D. Huntley, pp. 30–45. London: Ellis Horwood, 1991.

Waits, Bert K., and Franklin Demana. "Is Three Years Enough?" *Mathematics Teacher* 81 (January 1988): 11–14.

Part 4

A Synthesis Perspective

16

Looking Back, Looking Ahead

John Dossey

[E]fforts to correct historic imbalances in American higher education will require attention to questions of strategy. It is all well and good to suggest that common learning must assume a priority equal to uncommon learning, but we live in a world full of hierarchies that are built on credentials and standards, most of which reflect choices and values that encourage excellence at the expense of equality. In addition, we live in a world where it is impossible to separate the higher education systems from business, the professions, the military, government, and any number of other "systems." That makes it very difficult to initiate fundamental, systemic change. Given how difficult it is merely to change a college curriculum, which, relative to changing the balance among the purposes of higher education, is barely tinkering at the margins, it is hard even to imagine how one might go about the kind of truly radical change that would be necessary if liberal education were to be reconceived as a means to promote the problem-centered ways of thinking and to better combine those with discipline-based styles of thought.

—Ellen C. Lagemann

ALTHOUGH Lagemann (1997) was discussing curricular change in American higher education, she might as well have been discussing the issues related to instituting reform in school mathematics. Her reference to "strategy" signals that fundamental changes in education are now viewed as matters of public policy and open to a public debate. Curricular issues are no longer issues residing within the halls of the academy. The pressures that accompany change are not new to mathematics education, as questions of "balance" between social utility, disciplinary prerequisites, excellence, and equity have served as constraints in setting the direction for school mathematics over much of the last century (Jones 1970; Senk and Thompson 2003).

The mathematics curriculum projects described in this book illustrate the influences that the National Council of Teachers of Mathematics (NCTM) *Curriculum and Evaluation Standards for School Mathematics* (NCTM 1989) and the National Science Foundation (NSF) curriculum development projects have had on jump-starting reform efforts in mathematics education. These projects provide models for those interested in curriculum design and development to examine the role that differing materials have in supporting or inhibiting change in school mathematics curricula.

Looking Back

The origins of the reform movement that resulted in the NCTM *Curriculum and Evaluation Standards* and the NSF's request for proposals relative to the school mathematics curriculum were many. The mathematics and mathematics education communities realized that the then-present system of arithmetic-dominated elementary school programs and the sequence of compartmentalized secondary school courses that had long served as the basis for mathematics education in the United States were not providing students with an adequate foundation for either the workplace or higher education. This came to a flash point with the publication of *A Nation at Risk* by the National Commission on Excellence in Education (1983). The commission's findings echoed the NCTM's *An Agenda for Action* (1980) recommendations and the Conference Board of the Mathematical Sciences (CBMS) report on what was fundamental and what was not fundamental for school mathematics programs (CBMS 1983). The later reports of the 1986 National Assessment of Education Progress's mathematics assessment (Dossey et al. 1988) and the Second International Mathematics Study's findings (McKnight et al. 1987) added research findings and data supporting the need for curricular change.

As these calls for change began to mount, the National Council of Teachers of Mathematics was already working toward developing standards for what students should know and be able to do as a result of their study of school mathematics. The development of the plan for these recommendations and their subsequent release is well documented in the evaluative study of the process (McLeod et al. 1996). This, combined with a push from the nation's governors for grassroots change at the state level around the *Curriculum and Evaluation Standards,* signaled that change was seriously being considered.

As states changed their state assessment frameworks and curricular recommendations, publishers rushed to get out new versions of their texts, claiming that they "met the *Standards.*" Almost missed in this flurry of action across the country was the call by the NSF for proposals to create comprehensive sets of instructional materials for elementary, middle, and high school mathematics based on the NCTM *Curriculum and Evaluation Standards* and other sets of recommendations for change. It was this call and the successful proposers' products that led to many of the programs detailed in the earlier portions of this book.

Looking at the Projects' Designs and Development

An analysis of the reported design features of the fifteen projects indicates that they share, in general, many of the same goals regardless of the grade levels for which they were developed. Most projects report that they have focused on developing curricula that reflect a more applied approach to learning mathematics with a broader representation of geometry, data analysis and chance, and algebra and functions than in previous programs. Particular attention was paid in several programs to balancing the role of conceptual development with the honing of procedural fluency. Most projects report specific attempts to develop constructive paths to understanding prior to the establishment of some form of distributed procedural skill development.

Elementary school program design

A closer examination of the projects by grade-level bands exposes some differences in initial goals and design features. At the grades K–5 level, Trailblazers and Investigations began with a very modularized experiment/investigation approach linked to science. Everyday Mathematics and Think Math! started with rather full-blown versions of a curriculum plan. These two projects resulted from ongoing work that began long before the NSF request for proposals, whereas the former two projects were, in some ways, a product of the NSF funding itself. The projects group together to provide a contrast based on their content foci. Trailblazers and Everyday Mathematics had a broader focus on expanding students' view of mathematics, whereas Investigations and Think Math! gave a stronger emphasis to the development of number and operations.

Few of the programs initially created a complete set of design features. Although Trailblazers initially specified a distribution of activities allotted as 50 percent experiments, 25 percent problem solving, and 25 percent procedures, they abandoned it as they moved forward. Both Trailblazers and Investigations started with a much heavier focus on experiments and investigations as driving forces behind the organization of their curricula. Both eventually found they had to modify this because of time and teacher development issues. Perhaps the most interesting design issues emerging from the early work in the projects were the following:

- Developing students' understanding of procedures prior to practice with them and then seeing that the practice was distributed and tied to broader understanding of the concepts underlying the procedures

- Recognizing that representations provide both a basis for generalization and an avenue for entry to the concepts for less-able students. Capable students can transfer between representations and enlarge their understandings, whereas different representations offer different entry points for those who struggle to make sense of mathematical situations.

- Developing teachers professionally and providing that development in a fashion directly related to the instructional materials created for students. Trailblazers developed unit guides and fact guides and resource modules for local professional development. Everyday Mathematics followed much of the same course, developing related materials for professional development. They took the avenue of creating a corporation to help with the organization of the local delivery of the professional development. Investigations took the route of added teachers' notes, dialogue boxes with materials to guide classroom discourse, and essays about the content itself and interpreting students' work. Think Math! developed materials in such a way that the student materials provide professional development as the teachers work through it with their students.

It was disappointing to see that none of the projects focused on assessment development or gave technology a major role in their design plans. Although all recognized the need for assessment to change and saw the need for samples of students' work to be included in professional development, none made it a major design feature. A similar situation existed with technology. Although comments were made about technology's possible role in basic fact and operation-sense development, no project made it a major portion of the design decisions.

Middle school program design

At the middle school level, design assumed a more central role in project development. This may have been a result of the experience of the individuals involved or the nature of the presentations made in their papers. Connected Mathematics and Mathematics in Context both presented targeted lists of design features, and both MathScape and MATHThematics presented similar structures in prose. Although not all the projects presented identical lists, the main features common to the majority were the following:

- Organizing around important mathematical concepts, processes, and procedures
- Developing knowledge and skills from a problem-centered context
- Connecting ideas within units, across units, and across grade levels
- Expecting meaningful learning of procedures and distributed, motivated practice of them
- Developing communication and representation capabilities
- Creating links to a research based on learning and pedagogical practices
- Developing algebraic thinking based in proportional understandings

A strong trend of nurturing students' reasoning and generalization capabilities also ran through the projects' goals, as did the need for professional development and the building of individual students' motivation and ability to learn in groups or alone.

Connected Mathematics had the foundation and experiences emerging from the development of Michigan State University's Middle Grades Mathematics Project as an experiential launching pad, whereas Mathematics in Context made use of the work of the Freudenthal Institute's Realistic Mathematics Project from the Netherlands. MATHThematics and MathScape were created from the ground up, but they relied on strong advisory boards and a variety of other work being done in middle school mathematics at the time. All four projects relied on classroom teachers.

The design in each of the four projects was tied to presenting material to students in individual units that focus on a single topic or a set of tightly linked topics. This unit structure was used in each of the middle school projects except MATHThematics, which was published in hardcover format. The three unit-based projects used this format to give students a feeling of accomplishment as well as a sense of new beginnings. The weight of standard textbooks in backpacks was another concern. Regardless of how the materials were packaged, a study of their content development shows that the projects built in learning trajectories and coherence across units. Careful sequencing of lessons so that students saw how new material built on prior material was a major commitment of the writers. Further, the middle grades projects were driven by similar choices of content:

- Rational numbers and proportional reasoning
- Probability and data analysis
- Algebraic thinking and representations
- Geometric visualization and measurement

As such, all the programs are aimed at actively involving students in building bridges from the whole-number-dominated elementary school program to the mathematics of the secondary school. However, unlike traditional programs, the focus is on understanding and reasoning with the knowledge and skills in realistic settings. Students are confronted with problems whose solutions require deep understanding of the material, not direct applications of a formula, algorithm, fact, or classification of a shape.

Finally, most of the middle school projects attempted to address professional development in one fashion or another. The Connected Mathematics Project developed unit guides for teachers that included overviews and elaborations of the content contained in the student units. These were supplemented by classroom scenarios, questions, and communication prompts for classroom use. In addition, assessment suggestions were furnished along with access to a special Web site for further support. Mathematics in Context developed scope-and-sequence charts, pacing guides, and special assessment booklets for teachers to supplement the basic units. MathScape developed teacher-written essays on teaching the individual units and classroom tips on presenting individual lessons. MATHThematics developed classroom checklists and scales, discussion questions, and extended explorations for use in classrooms with its units.

Assessment and technology factors also worked into the planned designs for a couple of the middle school projects. Mathematics in Context and MATHThematics both developed special assessment products for use by teachers. The former materials focused on assessing students' mastery of a number of competencies, similar to those developed in Europe by Mogens Niss and his collaborators (Niss 2003) and Jan de Lange (1999). These approaches focus on students' capability to reproduce, connect and integrate, and generalize and reflect on their work. MATHThematics also produced classroom assessments and students' self-assessments for use with the instructional materials and developed a unit for teachers on portfolio assessment. Connected Mathematics provided teachers with four levels of assessment for each Investigation: Student Self-Assessment, Check-Ups, Partner Quizzes, and Unit Tests and Projects as well as sample assessments and students' work, discussing how they linked and made suggestions for teachers' own assessments. Technology also made an appearance in the middle school programs. The role varied from mere mention to MATHThematics's inclusion of specific exercises for both graphing calculators and spreadsheets. Tinkerplots was incorporated in Statistics units for MATHThematics and Connected Mathematics, and computer applets were developed for both these projects. Some were placed in the students' work, and others in teacher-based materials. In addition, more technology links were provided on a Web site.

High school program design

Design features were similar at the high school level to those viewed at the earlier levels. However, at this level, the recommendations for mathematical content and processes were even more explicit, in general, than they had been at the earlier grades. There was agreement on the basic content areas that appeared in the NCTM *Curriculum and Evaluation Standards,* but discrete mathematics and modeling were added in several of the programs as content focal areas. The curricular program that was perhaps most distinctive in this regard was Mathematics: Modeling Our World. In this program, modeling problems drove the units, and they were organized around major modeling categories, such as risk, fairness, and optimization. Core-Plus Mathematics integrated in its materials a significant amount of data analysis and probability, along with discrete mathematical modeling using vertex-edge graphs and recurrence relations.

Paralleling the content dimensions of design were the emphases placed on developing students' mathematical processes. Extending beyond problem solving and the remainder of the NCTM Process Standards, habits of mind mentioned were thinking recursively, optimizing, and abstracting and generalizing. All these are important facets of thinking mathematically. When combined with the process recommendations of the NCTM *Curriculum and Evaluation Standards* and the focus on similar habits of mind in the middle grades, this is an important feature of the design of secondary school materials if they are to achieve goals that differ from those of the instructional materials of the past few decades.

Only one of the secondary school programs made use of the design feature of replacement units. The Interactive Mathematics Project focused its instruction in six-to-eight-week units. However, several of the programs made explicit use of technology in their design work. The Interactive Mathematics Project included student exercises involving both graphing calculators and computers. The SIMMS Project reached further with materials involving Cabri geometry, a computer algebra system, and statistical and spreadsheet technology. Core-Plus Mathematics focused on graphing-calculator technology in its first edition but is moving on to include interactive data analysis and geometry software, spreadsheets, and a computer algebra system in its second-edition materials.

Addressing all students

Another common design feature that emerged from many of the reports was the emphasis on assuring that all students would make progress in mathematics, developing a feeling of self-confidence and self-assuredness. Accompanying this emphasis were attempts to create programs that would present mathematics as a dynamic field of study to be understood and applied, not a staid set of rules to be memorized. The design features focused on achieving these goals included experimentation with manipulatives and with technology, modeling in classroom sessions, and real-world data collection and study. Many of the programs discussed the role of collaborative learning in developing students' mathematical understanding and the furtherance of their capabilities to employ mathematical processes, individually and jointly. The NCTM Process Standards of problem solving, representation, communication, reasoning and proof, and connections received solid coverage in all the programs.

Program Development and Revisions

The review of the different programs' developmental plans, processes, and reflections offers insight into the genesis of these projects. The first fact that one notices is the value of the synergism that results from bringing teachers, mathematics educators, and mathematicians together on a common plane to work in the development of classroom materials. The reports of projects' development cycles detail the importance of the camaraderie that formed about the piloting and field tests and the work on revisions. A good part of this camaraderie came from the teachers' perceived ownership of the curriculum. Teachers involved were partners in the creation of the work, and it shows through in several of the reports.

Examining the different project reports as a whole, one notices commonalities and variations of development cycles into different formats. The project teams develop (D), revise (R), pilot (P), and field-test (F) for an individual grade level's set of materials. Across the fifteen projects, two of the projects used four-year cycles of D-R-P-R-F-R-F-R, with the first field test an experimental one and the second a national one. Another eight projects used a three-year cycle of D-R-P-R-F-R. The remaining five projects seemed to use variants of the D-P-R format, where there were substantial reactions and revisions taking place within the year of piloting. The actual authoring teams varied across the projects as well. In eleven of the projects, the writing teams were basically project-based, but several included former teachers, visiting teachers, or strong teacher reaction boards across the year. In the instances of Mathematics: Modeling our World and the University of Chicago School Mathematics Project—Secondary Component, drafts and revisions were substantially developed by teachers and mathematics educators working together in "summer camps" and at other scheduled meetings throughout the following year. The materials for MATHThematics and the SIMMS Project were written largely by teachers with support from mathematicians and mathematics educators.

However the materials were developed and whoever was included, it was clear that even the projects with the greatest amount of develop-test-revise cycles believed that they were rushed in getting the product out the door. Part of this may be because authors never want to let go of their curricular materials, but part of it reflects the extreme difficulty of writing materials with diverse forms of feedback. These projects strove to write, in general, for heterogeneous groups of students. This "upped the ante" with respect to writing something that a majority of reviewers would endorse in similar fashion. The fact that so many projects succeeded is the wonder!

Reactions and redirection

The reactions from the field both during and after the first published versions of the materials reached schools provided the projects with avenues for redirection. Common themes in the lessons learned included the following:

- Explorations took far longer than expected in classrooms. Analyses of the feedback in most instances showed that the projects had not honed the instructions down to the essentials and that they had included too many extraneous features in early explorations. Students need time to learn to filter extraneous features from central features. The response to this finding involved rewriting explorations to streamline them and, in some instances, cutting the number of explorations expected in a year.

- The lesson structure and teachers' notes did not clearly communicate to teachers what was central and what was not. When teachers struggle with new material and new formats, they often become lost in the details and revert to either a cover-it-all or skip-it-all approach. Both have serious consequences—time use and prerequisite knowledge, respectively. When this occurred, projects tended to respond with a better, perhaps more streamlined lesson plan and notes to the teachers on appropriate questions and activities based on classroom vignettes. The latter were sometimes made available by videotape for the teachers.

- At all levels, projects underestimated the amount of time and practice that it takes for students to develop procedural fluency with numerical operations and the manipulation of algebraic forms. Elementary school programs reacted by moving from strictly personal algorithms to the use of focus algorithms to bridge the span from understanding to the introduction of standard algorithms. They also developed fact and operation programs to supplement the core curriculum that still tried to focus on a broader view of mathematics than number and operation. At the secondary school level, more practice was added, especially for the mastery of symbolic manipulation skills in algebra. This necessitated the advancing of some topics in the curriculum and the inserting of review sections dealing with the maintenance of skills. Finally, across the board, more practice was built into the problems used to develop other concepts. Practice was distributed, rather than massed, in the hope of developing and shaping skills over a period of time, with the goal of student understanding and fluent use rather than memorization.

- Reading mathematics in context was a problem across the grades. Care was taken in revisions to lessen the amount of unnecessary reading while working to build the crucial reading skills that are necessary for doing mathematics. This is a continual balancing act and one that is made more difficult for students who speak English as a second language. Continued work remains to be done here.

- Project teams found that technology presented a double-edged sword. Too much technology at the upper end, even when supported and mastered, did not always match up with collegiate programs that did not allow students to use it. This created a problem when students were ill prepared to go it alone in situations without the technology backup. At the other end of the spectrum, curricular development projects can prepare excellent materials for use in technology-supported situations and then find that schools do not have the technology to offer that support. Different avenues have been constructed to cope with this situation. Core-Plus Mathematics has developed public-domain software; other programs have chosen to introduce technology gradually with successive editions; and still others have tried to say up front, "You need this [particular technology] if you are going to use our materials."

- At a different level, project directors found that in many instances they were not receiving adequate feedback on needed revisions from the field-test teachers. Part of this failure was due to teachers' work load, part due to teachers' inability to articulate the needed changes, and part due to the lack of direct contact time between the projects' staff members and teachers in field-test sites. Most of these problems were idiosyncratic to specific projects, but the sum of the number of mentions of this problem indicates that curriculum project teams needed to be aware of this in their planning for information gathering and revision cycles. The more successful projects seemed to develop better systems to collect the information and to establish closer contact with those using the materials. These procedures often involved logs, videos, special meetings, and several direct classroom visits.

- Another extracurricular problem that surfaced, especially in middle school settings, was the lack of mathematical knowledge on the part of teachers. This problem was addressed in many instances by developing teachers' notes or other added features to patch over the deficiencies. The Everyday Mathematics project did perhaps the most in creating a professional development arm of the project to assist local districts with their needs. They, as well as others, provided supplementary background publications for teachers that focused on both needed mathematics and pedagogical skills. Although a stopgap measure, these publications have the possibility of assisting districts while their teachers strive to teach through the materials, learning, it is hoped, along with their students and, one again hopes, with the aid of a mentor. Districts need to make commitments to help their teachers get appropriate backgrounds if they expect lasting change and growth to occur in their districts' classrooms.

When considered as a whole, it appears that the reform projects have shifted slightly back from their original designs, but they have maintained their philosophies in reacting to the comments they have received from the field in their field testing and from teachers and schools since commercial versions of the works have appeared. Several hints have arisen that suggest there may be difficulties ahead for projects to maintain their developmental cycles and philosophical positions as the publishing houses become more involved.

Looking Ahead

Historically, change in curricular programs in mathematics has tended to follow one of two models: revolution or evolution. In the 1960s we saw many instances of revolution: stark changes in the goals, content, and formats of instructional materials and the lack of teacher preparation to manage instruction with these changes. Many of the programs of the "new math" era crashed and burned. The residue of the changes they proposed, however, has gradually made its way into the curriculum, especially at the secondary school level—for instance, the reliance on properties, sets, and functions as organizational structures; the study of inequalities; and the presence of the precalculus and analysis course at the apex of the secondary school algebra curriculum. The evolutionary path often is viewed as one with a slow rate of change, as evidenced by successive editions of standard textbook series. This approach to change is based on tried methods and familiar content, but it is also one that both the mathematics and mathematics education communities and results of national and international assessments have identified as not serving historically to prepare our students to the level of modern expectations for students' performance in mathematics.

The change started by the NCTM *Curriculum and Evaluation Standards* (1989), by accompanying developments brought about through its renewal in *Principles and Standards for School Mathematics* (NCTM 2000), and by the materials produced by the NSF-sponsored mathematics curriculum projects provide another model for change—change guided by *Standards*-based expectations combined with commonly held principles for reform. Although the *Standards* have not yet reached the codification levels found in national curricula in other countries, steps toward providing even more guidance to schools, such as the NCTM's *Curriculum Focal Points for Prekindergarten through Grade 8: A Quest for Coherence* (2006), and the College Board's *Mathematics and Statistics: College Board Standards for College Success* (College Board 2006) furnish ample evidence, along with the convergence of state assessment standards, that there is a movement to establish more commonly accepted, grade-specific expectations for school mathematics in grades K–12. The reform curricula discussed in this volume, plus a few of the evolving traditionally based curricula, will serve as a basis for programs attempting to meet these new expectations. The design features discussed in this volume, plus the more inclusive and painstaking development projects augur for better products that are more closely aligned with teachers and classrooms.

Continued improvements will be guided by better evaluation plans based on the designs and questions suggested by the National Research Council's report *On Evaluating Curricular Effectiveness* (Confrey and Stohl 2004). Such studies will generate better and more targeted information for curriculum developers interested in improving the studied mathematics programs. The development of a solid evidentiary base for each program is mandatory, and the NSF and other funding agencies are already funding research efforts focused on such studies.

Lessons from the Reform Curricula

The possibilities for the coming years are at a tipping point at the moment of this writing. Increasing demands for annual growth and other statutory requirements imposed by the No Child Left Behind legislation have the prospect of placing a chokehold on progress and change. Assuming that change itself can bring these demands to a reasonable level and place the assessment and evaluation of students' growth on a firmer basis than achievement tests and large-scale state assessments, the possibilities for further growth and change are numerous.

Perhaps the largest and most daunting lesson learned from the implementation of the reform curricula is the need for professional development to accompany the rollout of new materials. This, perhaps more important than the materials themselves, may be the key to the changing of what happens in schoolrooms across America. Teachers not only need to see and experience the mathematical concepts and procedures contained in the materials but also need to do so in an atmosphere having the same discourse level as that in their classrooms. Can programs be built that will prepare teachers, both mathematically and pedagogically, to deal with the demands of teaching mathematics as the new curricula intended? The challenge associated with this is massive. It will need to encompass both preservice teacher education and professional development for in-service teachers. To be totally effective, it may need to even involve components for parents and administrators.

This need is not new and has historical roots in prior eras of reform. Across time from the University of Illinois Committee on School Mathematics (UICSM) in the 1950s to the implementation of graphing calculators in the classrooms during the past decade, successful implementations of innovative curricular materials in mathematics have been accompanied by carefully planned and classroom-linked professional development. Teachers are most likely to implement programs with fidelity when they have experienced the materials and activities themselves and have seen the results of their use with students. Reform programs in mathematics education must develop and institute such activities within preservice education programs and in the continuing professional development of practicing mathematics teachers. These programs cannot be one-day or one-week wonders but rather continuing, sequential programs that give teachers both content and pedagogical knowledge accompanied by the philosophical orientation that transforms their practice to align with the programs' curricular intentions. Some would suggest that really successful programs are those that involve teacher leadership and are held on local school sites—bringing the message of change home in a setting familiar to the teachers.

Closely tied to the topic of professional development is the development of lessons with students in the classroom. Compare the similarity in the instructional models in (*a*) Math Trailblazers: draw a picture, collect and organize data, graph the data, and analyze the data with regard to a leading question; (*b*) Connected Mathematics: launch, explore, and summarize; and (*c*) Core-Plus Mathematics: extension of launch, explore, and synthesize methods. At each level, one sees that the exploration is based on a problem-centered question. Students are immersed in a situation calling for innovation, reexpression or rerepresentation, interpretation, analysis or solution, reflection on the outcome, and communication of the findings as they relate to the initial problem. The question is how to move teachers from lecturing to questioning, from guiding to listening, and from testing to assessing. Much of this can be caught from colleagues in professional development. Some of it can be taught. But most of all, it probably needs to be experienced over and over until it seems natural and right to the teachers. Real change in instructional methods will come only when teachers themselves have learned in the same fashion and realize that it is a powerful way to help children—their students—to a brighter and more powerful understanding of mathematics. Along the way, teachers need to learn that the issue is not being afraid of not knowing an answer but instead of not being afraid to launch an investigation to find the answer. Teachers and students must become investigators together.

Usiskin and other contributors also warned of danger signals on the path ahead. Particularly insightful was the warning to avoid the oversimplification and over-application of the different reform factors. Students need to have the opportunity to learn in a variety of ways and to experience a variety of lesson formats. The application of the different design features should make sure that students have a variety of representations of instruction, assessment, practice opportunities, and technology experiences. It is only through such experiences that students can develop a palette of approaches to problem solving, procedures, and applications of technology.

Another important warning that came from the experiences of the developers, especially Max Bell and Andy Isaacs, was the danger of "creep and bloat." As programs move toward acceptance and commercial production, it is natural to try to make them fill all purposes for all states' curricular frameworks and the related mandated assessments. As a result, there is a little addition here, a little addition there, and soon the program looks less like it was intended and more like a program of disconnected pieces with too much to "cover" in the time available. A wise and trusted colleague, Frank Giordano, once said that "covering a curriculum" meant "turning pages to hide understanding." The creeping additions of this piece and that exercise to match some elusive target gradually leads to a bloating of the curriculum that forces teachers to race through material in order to cover it rather than experience it and reflect and apply it in meaningful mathematical activities. As the noted architect Ludwig Mies van der Rohe said, "Less is more."

Tied closely to this, the national community of mathematics educators and mathematicians has to work together with teachers and state officials in the modification of local, state, and national assessment systems to reflect the goals of reform-oriented programs. This transformation must not "throw out the baby with the bath water" while restructuring these programs. Radical change can destroy the base for long-term trend analyses related to material common to current and reform-oriented curricula. Care has to be given to carefully selecting the components of the assessment programs that are relevant to a balanced mathematics program that reflects both the development of conceptual knowledge and the creation and solidification of students' fluency with procedural and fact-based portions of the curriculum.

In 1990 the Mathematics Assessment of the National Assessment of Educational Progress was revised to include both open- and extended open-constructed response items, as well as to include items where technology was available and items where it was not. Manipulatives were available for some subtests but not for others. Thoughtful assessments need to be developed that will provide valuable information on what students can and cannot do under different conditions. Revised programs should give teachers feedback on students' thinking and problem-solving capabilities, as well as their procedural skills. Such assessments not only need to have a balance of different formats but also need to offer ample opportunities for students to give evidence of their capabilities to structure and complete projects reflecting extended problem-solving efforts. Assessments should also furnish evidence of students' capability to construct a justification, or proof, of a conjecture or offer a counterexample invalidating the conjecture. Good beginnings are available in the work of the Mathematics in Context materials, the evaluation model associated with the SIMMS materials, and the overall push to have students create products that can be viewed, questioned, and discussed with them.

One of the most important aspects of classroom assessment needed is a measure of a student's ability to learn mathematics as part of a group. The College Board's Pacesetter Mathematics program experimented with such a measure in the 1990s. In it, students' group-based mathematics abilities were estimated on the basis of a combination of a student's performance on an individualized test, a group project, and an individual assessment completed following a group-based planning session related to the topic of this last individual assessment. Such comprehensive pictures of a student's capabilities provide a deeper and richer picture of what a student can do, both individually and as part of a group. As students work more in collaborative settings in mathematics, teachers need tools to evaluate their students' learning in these situations—both to aid students and to adjust their instruction to maximize group effectiveness.

In several of the reports from individual projects' development activities, there were remarks detailing how the programs had been forced to lower their initial expectations for students' use of computer-based technology in mathematical learning situations. Many reported dropping back to expecting students to use only graphing calculators. As the projects continue to refine their materials, they need to continue to strive to reach their original expectations for students' explorations in spreadsheet, interactive geometry, data analysis, computer algebra, and modeling environments. Although graphing calculators are ubiquitous in the school mathematics classroom, they are not as common in the workplace or university classroom or laboratory. Students need to learn to do mathematics and statistics using standard software packages in computer-based environments as well as to master the use of graphing calculators.

Another issue tied to technology is the issue of a curriculum delivered through a CD. The increasing number of students being home schooled and others being taught in private or public schools through programs presented over computer networks raises the question about the degree to which these curricula can be viable candidates for such delivery systems. An even stronger question is whether they or any other program sharing similar goals would even believe such a delivery system is appropriate. Clearly, students in home schooling, in private schooling, or needing distance education should have the right to learn from the best materials available. However, the constructive, group-based learning experiences expected as part of the completion of the activities in these programs is such an integral part of the learning cycle that it is hard to see how it would be possible to offer comparable experiences to students in many of these situations. Moreover, how can one be assured that these students have access to discourse on the way to constructing their ideas? Careful research needs to be done prior to developing CD-delivered versions of these programs.

Tied to the need for discourse is the role of communication and critical reading in mathematics. These programs assume that students will engage in numerous activities that make use of their ability to read, speak, and write about their experiences in constructing an understanding of the mathematics they are studying. This expectation carries with it the implicit understanding that students are capable of reading mathematics in context. This is not the same thing as reading prose in context. The technical nature of mathematics and the use of words having both a prose and a technical meaning add to the load already brought by the use of symbol systems and numerical systems. Research needs to be conducted to further the earlier work of Kane, Byrne, and Hater (1974) in reading mathematics from text and in assisting teachers in helping their students develop the necessary skills. In a like manner, students need to learn to write in technical situations as well as in the more general prose-based settings they already experience in language arts.

Finally, mathematics educators, backed by the broader mathematical and statistical communities, need to play a more central role in policy work at the state and national levels. Decisions about the structure and guidance of school programs that have any effect on school mathematics programs, or students' associated learning and achievement, need to have mathematics and statistics educators at the table. Such policies can range today from legislation related to assessments, to the implementation of block scheduling, to the role discipline-related learning plays in the recertification of teachers. All these actions can place constraints on the environment in which mathematics learning takes place. When teachers of mathematics from all levels are not involved in these decisions, they, along with the students affected by the decisions, are apt to suffer.

There has been a great deal of change over the past two decades since the initiation of the development of the NCTM *Standards* documents. The mathematics education community has shown a cohesive, focused effort in moving the teaching and learning of mathematics forward. As economic challenges increase and funding for innovation lags, the mathematics education community must strive to support and further refine programs such as those described in this volume. Careful evaluations and related changes must be based on guidance from the National Academy of Science's reviews (Confrey and Stohl 2004; Kilpatrick, Swafford, and Findell 2001); on the evaluation programs associated with the programs themselves, substantiated with feedback on their own designs; and on continued listening to, and observing, the voices and actions of teachers and students who are actively engaged in learning mathematics using these, and like, materials.

References

College Board. *College Board Standards for College Success: Mathematics and Statistics.* New York: College Board, 2006.

Conference Board of the Mathematical Sciences. *The Mathematical Sciences Curriculum K–12: What Is Still Fundamental and What Is Not.* Washington, D.C.: Conference Board of the Mathematical Sciences, 1983.

Confrey, Jere, and Vicki Stohl, eds. *On Evaluating Curricular Effectiveness: Judging the Quality of K–12 Mathematics Programs.* Washington, D.C.: National Academies Press, 2004.

de Lange, Jan., and Freudenthal Institute Staff. "Framework for Classroom Assessment in Mathematics." Unpublished manuscript. Madison, Wis.: National Center for Improving Student Learning and Achievement in Mathematics and Science, 1999.

Dossey, John A., Ina V. S. Mullis, Mary M. Lindquist, and Donald L. Chambers. *The Mathematics Report Card—Are We Measuring Up?: Trends and Achievement Based on the 1986 National Assessment.* Princeton, N.J.: Educational Testing Service, 1988.

Jones, Phillip S., ed. *A History of Mathematics Education in the United States and Canada.* Thirty-second Yearbook of the National Council of Teachers of Mathematics (NCTM). Washington, D.C.: NCTM, 1970.

Kane, Robert, Mary Ann Byrne, and Mary Ann Hater. *Helping Children Read Mathematics.* New York: American Book Co., 1974.

Kilpatrick, Jeremy, Jane O. Swafford, and Bradford Findell, eds. *Adding It Up: Helping Children Learn Mathematics.* Washington, D.C.: National Academy Press, 2001.

Lagemann, Ellen C. "From Discipline-Based to Problem-Centered Learning." In *Education and Democracy: Re-imagining Liberal Learning in America,* edited by Robert Orrill, pp. 21–43. New York: College Entrance Examination Board, 1997.

McKnight, Curtis C., F. Joe Crosswhite, John A. Dossey, Edward Kifer, Jane O. Swafford, Kenneth J. Travers, and Thomas J. Cooney. *The Underachieving Curriculum: Assessing U.S. School Mathematics from an International Perspective.* Champaign, Ill.: Stipes Publishing, 1987.

McLeod, Douglas B., Robert E. Stake, Bonnie P. Schappelle, Melissa Mellissinos, and Mark J. Gierl. "Setting the Standards: NCTM's Role in the Reform of Mathematics Education." In *Bold Ventures, Vol. 3: Case Studies of U.S. Innovations in Mathematics Education,* edited by Senta A. Raizen and Edward D. Britton, pp. 13–132. Dordrecht, Netherlands: Kluwer Academic Publishers, 1996.

National Commission on Excellence in Education. *A Nation at Risk: The Imperative for Educational Reform.* Washington, D.C.: U.S. Government Printing Office, 1983.

National Council of Teachers of Mathematics (NCTM). *An Agenda for Action.* Reston, Va.: NCTM, 1980.

———. *Curriculum and Evaluation Standards for School Mathematics.* Reston, Va.: NCTM, 1989.

———. *Principles and Standards for School Mathematics.* Reston, Va.: NCTM, 2000.

———. *Curriculum Focal Points for Prekindergarten through Grade 8: A Quest for Coherence.* Reston, Va.: NCTM, 2006.

Niss, Mogens. "Quantitative Literacy and Mathematical Competencies." In *Quantitative Literacy: Why Literacy Matters for Schools and Colleges,* edited by Bernard L. Madison and Lynn Steen, pp. 215–20. Princeton, N.J.: National Council on Education and the Disciplines, 2003.

Senk, Sharon L., and Denisse R. Thompson, eds. *Standards-Based School Mathematics Curricula: What Are They? What Do Students Learn?* Mahwah, N.J.: Lawrence Erlbaum Associates, 2003.